THE EXISTENTIAL PLEASURES OF ENGINEERING

Also by Samuel C. Florman
ENGINEERING AND THE LIBERAL ARTS:
A Technologist's Guide to
History, Literature, Philosophy,
Art, and Music (1968)

THE EXISTENTIAL PLEASURES OF ENGINEERING

Samuel C. Florman

St. Martin's Press New York

St. Martin's Press, Inc., 175 Fifth Ave., New York, N.Y. 10010
Manufactured in the United States of America
Library of Congress Catalog Card Number: 75-9480

Library of Congress Cataloging in Publication Data

Florman, Samuel C
The existential pleasures of engineering.

Bibliography: p.
1. Technology—Philosophy. 2. Engineering.
I. Title.
T14.F56 601 75-9480

FOR DAVID AND JONATHAN

CONTENTS

PREFACE

Socrates said that the unconsidered life is not worth living. If the statement is valid, as I believe it is, then those of us who are engineers in the final quarter of the twentieth century are confronted with certain questions of compelling interest. What is the nature of the engineering experience in our time? What is it like to be an engineer at the moment that the profession has achieved unprecedented successes, and simultaneously is being accused of having brought our civilization to the brink of ruin?

Having posed these questions, it occurs to me that the answers are not without interest to all of us—engineers, would-be engineers, and concerned citizens in a world where engineering, for better or for worse, plays an increasingly important role.

This speculative essay is an attempt to find answers to these questions, and, as it turns out, to challenge some of the conventional answers that are being given by others both within the profession and without.

What I have written is brief enough to speak for itself and does not require any formal introduction. But I do think that it would be helpful to make a few comments in advance about three key words which will be recurring in what is to follow. The words are engineering, technology, and existentialism.

Engineers and writers about engineering are constantly lamenting the confusion that is supposed to prevail in the public mind about the meaning of the word *engineering.* I believe that such confusion is much less widespread than it used to be. For example, it is hardly necessary to be forever explaining that professional engineers do not wear striped caps and drive locomotives. It is generally recognized, I think, that engineering is "the art or science of making practical application of the knowledge of pure sciences." In other words, although engineers are not scientists, they study the sciences and use them to solve problems of practical interest, most typically by the process that we call creative design. Engineers are not mechanics, nor are they technicians. They are members of a profession. Although this profession has its roots in the earliest development of the human species, it only achieved recognition as a "learned profession" in the mid-nineteenth century, when scientific principles were first applied systematically to engineering problems, and when engineering schools and societies began to be established. In this book I have not attempted to describe in any detail what engineers *do.* Rather, my interest is in how engineers think and feel about what they do, and in the more general aspects of what it *means* to be an engineer.

Technology, like *engineering,* is a word that is constantly being defined and redefined. However, it is a word of such wide use and common understanding that I have chosen to take it pretty much as I find it, without bothering about linguistic or grammatical subtleties. *Technology* is a broader and more comprehensive term than *engineering.* Yet there is little question that engineers are the educated professionals who play a dominant role in technological development. Clearly the growing intellectual movement of "antitechnology" directs its hostility against the engineer as the archetypical technologist. Where I thought it appropriate, I have not hesitated to use the words *technologist* and *engineer* interchangeably.

Existentialism is a word that hardly anyone had heard of before it came into vogue after World War II. In this sense, existentialism is the newest of philosophies. Paradoxically, it is also the oldest of philosophies. This is so because its basic tenet is that existential thought precedes all formal philosophical deliberation. The existentialist is man saying to himself, "I exist! The way I *feel* about this existence of mine means more to me than any theory in the world." Existentialism, as a philosophy, has been argued and embellished by the many different thinkers who have been characterized as existentialists. But I have restricted the use of the term to its most essential meaning, which I take to have two aspects: (1) rejection of dogma—particularly scientific dogma; and (2) reliance on the passions, impulses, urges, and intuitions that are the basic ground of our personal existence.

At first glance, engineering and existentialism appear to have nothing to do with each other. The engineer uses the logic of science to achieve practical results. The existentialist is guided by the promptings of his heart, which, as Pascal said, has its reasons that reason cannot know. The existentialist most typically sees the engineer as an antagonist whose analytical methods and pragmatic approach to life are said to be desensitizing and soul deadening—in a word, antiexistential. To show that this adversary relationship is based on a misapprehension of the nature of the engineering experience is—as can be surmised from the title—a principal objective of this book.

PART 1

1. "WHAT WAS TROY TO THIS?"

In May 1902 the fifty-year-old American Society of Civil Engineers held its annual convention in Washington, D.C. Robert Moore, the newly elected president, gave a welcoming address entitled, "The Engineer of the Twentieth Century." He began by eulogizing the engineers of the past for making human life "not only longer, but richer and better worth living." Then he acclaimed the achievements of his contemporaries and fellow members. Finally he warmed to his chosen topic, the engineer of the coming era:

> And in the future, even more than in the present, will the secrets of power be in his keeping, and more and more will he be a leader and benefactor of men. That his place in the esteem of his fellows and of the world will keep pace with his growing capacity and widening achievement is as certain as that effect will follow cause.

What a flush of pleasure they must have felt, those engineers of 1902, to hear themselves described as benefactors of mankind. What a quickening of the pulse there must have been as they listened to their leader predict success and glory for them in the years ahead. Doubtless they sat quietly, looking solemn in their starched collars and frock coats, the way we see them in faded photographs. But beneath those sedate façades they could not have helped but feel the stirrings of a fierce joy.

To be an engineer in 1902, or at any time between 1850 and 1950, was to be a participant in a great adventure, a leader in a great crusade. Technology, as everyone could see, was making miraculous advances, and, as a natural consequence, the prospects for mankind were becoming increasingly bright.

Every few months, it seemed, some new technological marvel was unveiled and greeted with wild public enthusiasm. There were marvels of transportation—trains, ocean liners, trolley cars, subways, automobiles, dirigibles, and airplanes; marvels of communication—telegraph, telephone, phonograph, movies, radio, and television; marvels of construction—bridges, tunnels, dams, and skyscrapers; miraculous new sources of power—steam engines, gasoline engines, Diesel engines, electric dynamos; wondrous new materials—steel, petroleum, aluminum, rayon, and plastics; machinery to save labor and expand production—reapers, looms, presses, derricks, and lathes; and, of course, the innumerable conveniences of daily life that provided perhaps the biggest thrills of all—sewing machines, toilets, typewriters, bicycles, cameras, watches, electric lights, refrigerators, air conditioners, and so forth.

The completion of sizable technological undertakings was marked with celebrations fitting for an armistice or a coronation. Twenty-five thousand people paraded around the streets of New York for sixteen hours when the first cablegram was sent from America to Europe. At the moment that the Union Pacific and the Central Pacific railways were joined at Promontory in 1869, all over America bells were rung, cannon were fired, and bonfires were lit. That same year a glittering array of royalty and a procession of flag-bedecked ships marked the inauguration of the Suez Canal. As late as 1937, the opening of the Golden Gate Bridge was celebrated by a "Fiesta" attended by 200,000 persons, followed by a week of pageants and other festivities.

At the great international exhibitions, starting with London's Crystal Palace in 1851, technology was literally idolized. Six million people visited the Crystal Palace in Hyde Park to gape at the miraculous new machines on display. At the 1876 Centennial Exhibition in Philadelphia, in the glass and iron Machine Hall, the Corliss steam engine dominated the scene like a gargantuan icon. Alexandre Eiffel's one-thousand-foot tower was the sensation of the 1889 Universal Exposition in Paris, attended by more than thirty-two million visitors. Again in Paris, in 1900, Henry Adams tells of how he "haunted" the machinery exhibits, mesmerized particularly by the "occult mechanism" of the dynamo. Every few years—at St. Louis (1904), Brussels (1910), San Francisco (1915), Wembley (1924), Philadelphia (1926), Chicago (1933), New York (1939)—the engineers and inventors presented their new magic show to a dazzled and appreciative public.

In the late 1800s engineers began to appear regularly as heroes in novels and short stories. Rudyard Kipling's *Bridge-Builders* were depicted as robust creators of the British empire. August Strindberg eulogized in fiction the builders of the St. Gotthard Tunnel. The engineer hero of Zane Grey's *The U.P. Trail* was "wild for adventure, keen for achievement, eager, ardent, bronze-faced, and keen-eyed," a man who "had been seized by the spirit of some great thing to be." A whole series of best sellers glorified the American civil engineer of the early 1900s—*The Iron Trail, End of Steel, The Trail of the Lonesome Pine, Whispering Smith, The Fight on the Standing Stone, The Fire Bringers, Empire Builders.* When one of this genre, *The Winning of Barbara Worth,* was made into a movie, Ronald Colman played the part of the handsome engineer who gets the girl. Jules Verne's more than fifty fantastic tales were perennial favorites; so were the futuristic visions of H. G. Wells. As for the poets, in 1855 Walt Whitman was "Singing the great achievements of the present,/Singing the strong light works of engineers," a hymn of affirmation that Carl Sandburg was to echo sixty years later. On the occasion of the completion of the transcontinental railway, Joaquin Miller, a popular poet of the day, wrote, "There is more poetry in the railway that crosses the Continent than in all the history of the Trojan War." Robert Louis Stevenson, in his journal, *Across the Plains,* used the same epic image to describe the building of the railroad. "If it be romance," he wrote, "if it be contrast, if it be heroism we require, what was Troy to this?"

A few querulous voices were raised in alarm against the coming of the new machines and deplored the worship of material progress. Thoreau comes first to mind, of course, then Samuel Butler, Matthew Arnold, Nathaniel Hawthorne, and others, up to Aldous Huxley whose *Brave New World* appeared in 1932. Some of those writers who most admired technology—Whitman, Henry Adams, and H. G. Wells, for example—also feared it greatly. Charlie Chaplin's *Modern Times* expressed a wry but deep-felt protest. Yet the doubts and objections of a handful of artists and intellectuals could not stem the tides of public opinion. The conventional wisdom was that technological progress brought with it real progress—good progress—for all of humanity, and that the men responsible for this progress had reason to consider themselves heroes.

The years between 1850 and 1950 were indeed good years for engineering, "the Golden Age of Engineering," one is tempted to call them, in today's language of nostalgia. Before 1850 there had been many fine engineers and many outstanding engineering works. But engineering itself had been rather a craft than a profession, relying more on common sense and time-honored experience than on the application of scientific principles, and lacking those essentials of true professionalism—professional schools and professional societies. After 1950, as we shall see, engineering entered into a dark age of criticism and self-doubt. But during the hundred years between, the profession flourished. Its mighty schools and societies proliferated. And as its members grew in numbers (from two thousand to more than half a million in the United States), they also grew in prestige, power, accomplishment, and self-satisfaction.

The self-satisfaction came from many different sources. First of all, there was the elementary pleasure of solving technical problems and successfully completing constructive projects. This pleasure was as old as the human race. What was new about engineers as they started to develop as a profession was the delight they took in thinking of themselves as saviors of mankind. Since the lot of the common man had traditionally been one of unrelenting hardship, engineers looked upon their works as man's "redeemer from despairing drudgery and burdensome labor."[1] Once the common man was released from drudgery, the engineers reasoned, he would inevitably become educated, cultured and ennobled, and this improvement in the race would also be to the credit of the engineering profession. Improved human beings, of course, would have to be happier human beings.

Next, elevation of the common man would tend to make all men more nearly equal, thus making the engineer an agent in the realization of the democratic dream, "an apostle of democracy," as one engineer orator put it in 1905.[2] Comfort, leisure, and equality would imbue men with confidence in themselves and in the objectives of their society, making for the growth of patriotism and domestic tranquillity. Jealousies and class hostilities would diminish, a sense of brotherhood would spread, and the cause of peace would be served.

The engineer's works would also contribute to brotherhood by literally bringing men closer together. In the earliest days of the American republic, both Jefferson and Hamilton had remarked on how roads and canals would serve this end. Through the years, each advance in transportation and communication evoked new commentary on the theme. Walt Whitman rhapsodized:

> Lo, soul, seest thou not God's purpose from the first?
> The earth to be spann'd, connected by network,
> The races, neighbors, to marry and be given in marriage,
> The oceans to be cross'd, the distant brought near,
> The lands to be welded together.

As the Panama Canal neared completion, poet Percy MacKaye exulted over this wondrous work:

> Where the tribes of man are led toward peace
> By the prophet-engineer.

But all of this was only a small part—the most obvious part—of the satisfaction that engineers found in their work. They felt that they were improving the world, not only by their deeds, but also by their *way of thinking*. If engineers could solve problems by being open-minded and free of preconceptions and prejudices—by applying scientific methods—could not all men learn to think in this mode, and then would not ignorance, superstition and bigotry vanish? "We are the priests of the new epoch," an engineering leader told his colleagues in 1895, "without superstitions."[3]

Ever since the days of the eighteenth-century Enlightenment, men had been tantalized by the thought that social and moral truths existed, just like scientific truths, and could be discovered by the application of "right reason." This idea appealed mightily to engineers as their profession came of age. They fancied themselves the group best qualified to show the way in this great enterprise. Were they not, after all, the appliers of scientific method to practical problems?

"There is evil and plenty of it, the world over," wrote John Augustus Roebling, designer of the Brooklyn Bridge, "but all this evil may be traced in its origin to man's transgression of the laws of his own being." If the laws of man's being were discoverable, like the laws of thermodynamics, then indeed the blueprint for Utopia might be close at hand. In his native Germany, Roebling had been a student of the philosopher, Hegel. He brought with him to America the Hegelian concept that man might eventually free himself from the irrationalities of history by mastering nature. Not all engineers studied Hegel, of course, but many were influenced by Herbert Spencer (himself an engineer), who, during the 1880s and 1890s, expounded his version of Darwinism. In Darwin's theory of evolution, Spencer saw proof that man and society were governed by scientific laws, just as nature was. The first president of the American Society of Mechanical Engineers cited Spencer in declaring that the engineering profession should concern itself with general problems of politics and economics, since they might well prove susceptible to engineering solutions.

Frederick W. Taylor, the mechanical engineer who, around the turn of the century, "discovered" the rationalization of work (time and motion studies, quotas, unit costs—what came to be called efficiency engineering) considered himself a prophet whose "scientific principles" would settle all social conflicts. The principles of scientific management, he averred, "can be applied with equal force to all social activities: to the management of our homes; the management of our farms; the management of the business of our tradesmen large and small; of our churches, our philanthropic institutions, our universities, and our governmental departments." Henry L. Gantt, a disciple of Taylor's, went so far as to form an organization in 1916 called the New Machine. Society, declared Gantt, "must accord leadership to him who knows what to do and how to do it for the benefit of the community. This man is the engineer."

For those engineers who were neither formal philosophers nor social visionaries, there was always support to be found in that most venerated of sources—the Bible. For had not God given the earth to man and ordered him to "subdue it"? Many engineers saw their work as the carrying out of the Christian mission of subduing the earth for the benefit of man and for the greater glory of God.

In short, engineers of the Golden Age were not at a loss for intellectual, philosophical, and spiritual gratifications to go along with their often not inconsiderable tangible rewards.

There were a few frustrations, to be sure. Although they were generally admired and occasionally lionized, engineers never seemed to be satisfied with the "status" accorded them. There was a social basis for this: in Europe, engineering had been a career traditionally shunned by the upper classes. (When Herbert Hoover told a lady he was an engineer, she replied, "Why, I thought you were a gentleman!") There was a professional basis as well: engineering was the youngest of the professions, with none of the established traditions of medicine, divinity, and law. And engineers, jealous of their new identity, were repeatedly irked by being confused with scientists, or rather for having scientists get all the credit for engineering accomplishments.

But the greatest irritant by far was the domination of so many engineering activities by the business community. For all their declarations of professional independence and moral purity, engineers again and again found themselves subservient to financiers and businessmen. Indeed, serious cleavages developed between those engineers who pledged allegiance to the business community, for whom so many of them worked, and those who attempted to build an aloof and independent profession. These cleavages were largely responsible for the splintering of the profession through the proliferation of diverse engineering societies.

Another embarrassment was the failure of the engineer's proclaimed social programs to really take hold. Compared to all the talk of applying engineering methods to social problems, results were exceedingly meager. Henry Gantt's organization, the New Machine, came to naught, as did all similar ventures. During the 1920s Thorstein Veblen aroused some public interest in "technocracy," a proposed takeover of power by "a soviet of technicians." But when the idea was revived briefly in 1932, most engineers considered it to be a maverick movement controlled by left-wing impostors.

These few annoyances and embarrassments paled, however, in contrast to the real accomplishments, the popular acclaim, and the promising dreams to which the engineering profession could point. Whatever the engineering profession had not achieved between 1850 and 1950, it certainly expected to achieve in the very near future. During World War II, the development of radar, sonar, and other techniques promised new engineering marvels for the postwar years. Even the horror of Hiroshima was quickly forgotten in talk of atomic power for peaceful purposes. "The engineer," wrote a noted historian of the profession, "has faith in his work and in the ultimate beneficence of the forces he serves."[4] "It is a great profession," reminisced Herbert Hoover as the twentieth century reached its midpoint. Few engineers of the previous hundred years would have disagreed.

If we had asked an engineer of the Golden Age whether he found engineering existentially satisfying, he probably would not have understood the question. For the word *existential* achieved currency only with the popularity of the work of the French existentialist philosophers after World War II. But engineers did, I believe, find their work thrilling in a deep-down, elemental way that we think of when the word *existential* is used today. They felt fulfilled as men. They felt a part of the flow of history. They loved their work and believed it was inherently good.

At least, that is how it appears to me as I look back upon that age. And that is how it appeared to me as a boy growing up in the 1930s thinking about becoming an engineer. I remember well those old newsreels showing the dedication of yet another TVA dam. Instead of the floods and dust bowls of which we had seen so much, here were peaceful scenes of tamed rivers and humming transmission wires. We applauded F.D.R. as he promised more of such wonders in the years ahead. When I visited the General Motors Futurama Exhibit at the 1939 New York World's Fair, I believed that I was truly looking at "The World of Tomorrow." It promised to be a better world, too. It would have to be, with its superhighways, its sleek cars, and its sparkling cities. When I graduated as an engineer in the midst of World War II, the dreams of the engineering profession were still intact. The war was a temporary nastiness, after which the building of a better world would be resumed. In the Seabees—Navy construction battalions led by civil engineer officers—the proud boast was "Can Do!" "The difficult we do immediately," we used to say. "The impossible takes a little longer."

2. DECLINE AND FALL

It can be contended that the Golden Age of engineering was only beginning, rather than ending, in 1950. Since that time we have seen the growth of electronics, lasers, systems analysis, atomic power, computers, and a hundred other disciplines that make all previous engineering seem like mere tinkering. We have ventured into space, and miracle of miracles, landed a man on the moon.

Beyond question, since 1950 the most marvelous engineering feats have been performed each day, even more marvelous than they sometimes appear to our limited understanding and to our rather jaded sense of wonder. But how is it possible to appreciate these achievements when the very foundations of the profession are being attacked and appear to be crumbling? What satisfaction can there be in recording yet another technical triumph if the ultimate value of engineering as an activity has been brought into question? It is being said that engineering, no matter how clever, is destructive. It is being said that engineering, no matter how well-intentioned, is pernicious. Engineers are being called charlatans, fools, and devils. And such things are not being said by a single eccentric philosopher sitting by Walden Pond, but by myriads of people in every walk of life. Even engineers, to judge by their journals, have become uncertain, self-critical, and defensive. It should be apparent that engineering's Golden Age ended abruptly about 1950, and that the profession, for all its continuing technical achievements, finds itself at the present time in a Dark Age of the spirit.

If I had to choose a moment to mark the beginning of the downfall of the engineer, I would make it January 31, 1950, the day President Truman announced that work would begin on the development of a hydrogen bomb. The decision to create an "ultimate" weapon, more horrible even than those already existing, made evident, as nothing else could, the latent destructiveness of technology. A few days after the announcement, Albert Einstein wrote:

> Radioactive poisoning of the atmosphere and hence annihilation of any life on earth has been brought within the range of technical possibilities. . . . In the end, there beckons more and more clearly general annihilation.

At that point in history the average man could not help but begin to wonder about the technology in which he had put so much faith. There was some talk about the sins of science, but it was really the *application* of science through technology, or engineering, that suddenly seemed malevolent. When the Korean War broke out in June, vague anxiety turned to real concern, and even to panic. There was a brisk business in fallout shelters. Before the end of 1952 the United States exploded its first H-bomb, and the chilling statistics of its destructiveness began to circulate. Nine months later, the Russians had an H-bomb of their own. The Korean War ended without Armageddon, but it soon became apparent that war was not the only danger humanity had to fear from the atom. In 1954 an American test at Bikini atoll showered a boatload of Japanese fishermen with lethal radioactive coral dust. Then there were stories of strontium 90 appearing in milk, and of other health hazards resulting from atmospheric testing of nuclear weapons.

Fears of nuclear holocaust and radioactive poisoning eventually receded, only to be replaced by other apprehensions about technology. A best seller of 1960 was Vance Packard's book, *The Waste Makers.* With alarming statistics, Packard showed how Americans were squandering their limited resources. He accused engineers of compromising their ethical standards, and betraying their fellow citizens, by participating in the planned obsolescence of consumer products. Two years later Rachel Carson's *Silent Spring* burst upon the scene. We are destroying the natural world around us, announced Miss Carson, and unwittingly poisoning ourselves as well. By the introduction of chemicals into the food chain, we are planting the seeds of disease, madness, and death for ourselves and for our children.

Scarcely had the idea of poisoned food and water sunk in, when people were told that the atmosphere was being contaminated. The cars and trucks of which we were so proud, and the power plants and factories on which we were so dependent, were apparently drowning our communities in soot and dangerous gases. The infamous London smog of 1952, which was held responsible for 4,000 deaths, apparently was not a freak occurrence, but more likely a harbinger of things to come. "The Coming Struggle to Breathe" was the graphic title of a 1964 article in *The Saturday Review*.

1965 was the year of Ralph Nader's *Unsafe at Any Speed*, an indictment that sent shock waves through the engineering community. Nader angrily accused automotive engineers of failing to subscribe to any professional code of ethics. They ignore the safety of the public, he declared, and serve only the selfish commercial interests of their employers.

By this time the Vietnam war had grown from a nuisance to a major tragedy, and each evening's television newscast confronted the viewing public with such technological creations as napalm and defoliating chemicals. What this did for the "image" of the technologist needs no amplification.

As might be expected, an antitechnology backlash began to develop. In the summer of 1969, at the very time that men were walking on the moon for the first time, a book appeared in the stores with the eye-catching title, *America the Raped—The Engineering Mentality and the Devastation of a Continent.*

Public concern reached a climax in April 1970 with the celebration of Earth Week. All across the nation people expressed their fear and outrage over the environmental crisis. Senator Vance Hartke, speaking to a college audience, placed the blame where most people seemed to think it belonged: "A runaway technology, whose only law is profit, has for years poisoned our air, ravaged our soil, stripped our forests bare, and corrupted our water resources."

Barry Commoner's *The Closing Circle*, published in 1971, seemed to wrap it all up. We are, said Commoner, on the verge of destroying our air, our soil, our wildlife, our water, our foodstuffs—the entire ecosphere—and inevitably ourselves. Why is this happening?

> The over-all evidence seems clear. The chief reason for the environmental crisis that has engulfed the United States in recent years is the sweeping transformation of productive technology since World War II.

Commoner's facts were persuasive and his logic compelling. His style was calm. But his message was far from calming. "Underneath the gentle exposition," said *Newsweek*'s reviewer, "can be heard the urgent scream."

Nor was disenchantment with technology restricted to rational consideration of its dangers. Disenchantment overflowed into every corner of society, and evidenced itself in every important social change. When the Beat Generation emerged in the mid-fifties, a disdain for technology was its rallying cry. When the hippies wore flowers in their hair, it was the materialism of industrial production that they particularly rejected. The 1964 student disturbances at Berkeley indicated that a disaffection of almost unfathomable dimension had set in. A leader of the demonstrators called upon his followers to stop "The Machine." One student carried a placard that captured the nation's attention: "I AM A HUMAN BEING—PLEASE DO NOT FOLD, SPINDLE OR MUTILATE." When students rioted in Paris a few years later, the message was less whimsical. "Death to the Technocrats!" read a slogan painted on a wall.

The new rock music gave voice to a melancholy protest. Cat Stevens's verse was representative:

> Well I think it's fine building Jumbo planes, or taking a ride on a cosmic train, switch on summer from a slot machine, yes, get what you want to, if you want, 'cause you can get anything.
>
> I know we've come a long way, we're changing day to day, but tell me, where d' th' ch'ldr'n play?[1]

This is the language of the youthful counterculture. But a distaste for technology was not limited to the young. Sophisticated nightclub audiences laughed nervously as entertainer Tom Lehrer sang:

> When the air becomes uranious
> We will all go simultaneous

and then, with a special jab at the engineer:

> When the rockets go up who cares where they come down?
> "That's not my department," says Wernher Von Braun.[2]

The world of the high arts has always had running through it a strain of hostility toward technology. But since 1950, the fastidiousness of the few has become the revulsion of the many. In the postwar European cinema we start to see factories, office towers, and half-finished apartment complexes standing ominously in the background, at once a lamentation and a reproach. In Michelangelo Antonioni's film, *Red Desert*, the heroine wanders, lost and forlorn, through an oil refinery. Jean-Luc Godard's *Weekend* dwells with horrible fascination on an incredible series of automobile accidents.

In novels of the fifties and sixties, the protagonist is usually found struggling for liberation from an oppressive technological environment. "Rabbitt" Angstrom of John Updike's *Rabbitt, Run*, is typical. Returning reluctantly to his home town after an aimless all-night automobile drive to nowhere, he sees it "as a treeless waste of industry, shoe factories and bottling plants and company parking lots and knitting mills converted to electronics parts and elephantine gas tanks lifting above trash-filled swampland . . ." As the novel ends, Rabbitt is running, but there is no escape from his world of drabness and frustration.

Some novelists have looked into the future and shared with us their unsettling visions. Kurt Vonnegut, in *Player Piano*, has one of the leaders of a doomed revolt against the established technocracy say plaintively:

> I propose that men and women be returned to work as controllers of machines, and that the control of people by machines be curtailed. I propose, further, that the effects of changes in technology and organization on life patterns be taken into careful consideration, and that the changes be withheld or introduced on the basis of this consideration.

In *A Clockwork Orange*, Anthony Burgess forces us to take a chilling glimpse into a world peopled by aimless hoodlums, and worse, by technologists able to manipulate human beings so as to deprive them of moral choice. Even in the field of science fiction, apprehension is replacing the optimism of earlier times.

All about us the sense of disenchantment with technology appears to be growing. No one who is interested in the engineering profession—least of all those of us who are engineers—can ignore the fall of the engineer from the dizzying heights he once occupied. Sometimes it is easy to forget what has happened, since we engineers are all busy, our successes are many, and the majority of the people still seem to treat us as pretty respectable citizens. But there can be no denying that, with the coming of the environmental crisis, our relationship to society has changed. We cannot—should not—pretend that it has not happened, or that a hundred space spectaculars can restore things to what they were.

The self-satisfaction of our professional forebears now haunts us like a bittersweet memory of vanished youth. The pleasure in technical achievement is still there, of course. But what of the delight engineers used to take in their role as saviors of mankind? Clearly we have saved nobody—or, more precisely, those we have saved are now endangered by poisons and other hazards that we have created. Where men have been released from drudgery, they do not appear to have become superior human beings. Hardly anybody seems to think that they are more content. Anxiety and alienation are the watchwords of the day, as if material comforts made life worse, rather than better. Dreams of the equality, brotherhood, and peace which were to follow in the wake of engineering triumphs have also proved to be vain illusions.

As for the ultimate hope, that the engineer's rational thinking would show the way toward solutions of society's problems, the unanticipated events of each incredible, tumultuous day demonstrate convincingly what a naïve conceit that was. B. F. Skinner may be correct in his idea that men can be remolded by manipulation of their environment. But to apply his theories on a grand scale appears to be impractical, to say nothing of undesirable. So for the foreseeable future man's irrationality—which engineering philosophers have seen fit to ignore—will have to be taken very seriously into account. In retrospect, the ideals and dreams of engineering's Golden Age seem foolish and immature.

Each engineer old enough to remember what it was like "before," must have his own private experiences which have made real to him this devastating change. For me, it has happened when I find *The New York Times* denouncing plans to build power-producing dams on the Upper Missouri as "an act of vandalism" and "a desecration," and when I see books about the Army Corps of Engineers entitled *The River Killers* and *Dams and Other Disasters.* Apparently my youthful dream of a national paradise modeled after the TVA has become something of a national nightmare. As for that other vision of my younger years, the 1939 Futurama exhibit, it is difficult to think about it now without wincing. John Brooks's comment, in *The Great Leap* (1968), is not untypical:

> . . . many of its prophecies of heaven have become facts; the only trouble is that now that they are here they look more like the lineaments of Hell.

The "Can Do!" boast of World War II years now seems as pathetic as a battered landing barge rusting on a Normandy beach.

3.
CONSCIENCE, ERROR AND RESPONSIBILITY

In the wake of the environmental crisis, a reappraisal of the engineer's role in society would seem to be in order. This requires sober reflection and earnest intellectual effort. Instead of this, the profession seems suddenly to be embarked on a frenetic evangelical crusade. People who should know better—many engineers among them—are suggesting that most of our problems would go away if only engineers would become *more moral.* The environmental crisis would never have occurred, so the reasoning goes, if engineers had only given warning and refused to work on evil projects. They did nothing to prevent the crisis because they were greedy and ambitious, or inexcusably obtuse, or, at best, fearful of their industrial employers. Editorials and sermons now call upon the engineer to develop a conscience so that the world can be saved.

It is perhaps not surprising that engineers themselves are being caught up in this quasi-religious excitement. If repentence is all that is required to restore us to power and grace, then of course we will repent. The messianic complex dies hard.

Engineering journals are full of articles and letters to the editor suggesting that the profession develop new and improved ethical standards. Conferences are being held entitled "The Social Responsibility of Engineers." At one such conference, sponsored by the New York Academy of Sciences, the chairman hailed the coming of "a new consciousness" and "new concepts of professionalism." At another, an engineering student berated his elders, asking "how many times a man can sell out his moral convictions and still think of himself as a man."[1]

What an ironic turn of events! For if ever there was a group dedicated to—obsessed with—morality, conscience, and social responsibility, it has been the engineering profession. Practically every description of the practice of engineering has stressed the concept of service to humanity. Thomas Tredgold's classic definition dates from 1828: "Engineering is the art of directing the great sources of power in nature for the use and convenience of man." The definitions have been pouring forth ever since, most of them saying the same thing with just a few words changed here and there: ". . . for the benefit of the human race." (Henry G. Stott, 1907); ". . . for the general benefit of mankind." (S. E. Lindsay, 1920); ". . . for the good of humanity." (R. E. Hellmund, 1929); "applying . . . to the needs of mankind." (Vannevar Bush, 1939). Codes of Ethics have flowed almost as copiously as definitions, and these have also stressed devotion to the needs of the community at large. There was for a time resistence among some engineers to promulgation of formal ethical codes, it being said that a professional gentleman's conscience was the best guide in such matters. But by 1950—just when engineering's time of troubles was about to begin—most engineering organizations had formally endorsed the Fundamental Principles of Engineering Ethics drafted by the Engineers Council for Professional Development. This document stated unequivocally that the engineer "will use his knowledge and skill for the advancement of human welfare." The annals of the various professional societies bulge with speeches and papers emphasizing that upright character and concern for the public are the very essence of engineering. Most engineers have been good men devoted to good works. One shudders to think what the world would be like if this had not been so.

But long ago engineers discovered that fine sentiments and professional prestige were ineffective in curbing the excesses of technological development. Entrepreneurs were not easily dissuaded from seeking profit wherever it was to be found. The engineer's power was usually limited to a refusal to endorse a given plan, or to the tender of his resignation. This was not likely to stop a robber baron in his tracks. But the early American engineers were not unduly surprised by this state of affairs. Although they were men of conscience, they did not assume that the world could be ruled by conscience alone. Civilized men had long recognized that laws and regulations, mutually agreed upon, are the only sound protection for society against the self-interest of each of us. The founding fathers of the Constitution considered this as a given. James Madison asserted that men needed governing because they were not angels. Even Thomas Jefferson, that great believer in popular democracy, said, "In questions of power, then, let no more be heard of confidence in man, but bind him down by the claims of the Constitution."

Consequently, many engineers who were interested in the welfare of society, and who found themselves frustrated in industry, enlisted in the service of the only social institution powerful enough to effectively represent the public—the governmental regulatory agency.

Municipal engineers of a century ago struggled to protect the populace against hazards that most of us have forgotten ever existed. Charles F. Chandler, who served as advisor to the New York Metropolitan Board of Health, waged successful campaigns against watered milk, adulterated and explosive kerosene, poisonous cosmetics, the obnoxious smell from gas manufacturing, contamination from manure, and the spread of disease from filthy conditions around the meat markets. He helped to institute plumbing codes and municipal inspection of tenements. In all of these endeavors he was vigorously opposed by powerful business interests and corrupt politicians. Many other engineers have labored as Chandler did, and are laboring still, to protect the public interest.

The current proponents of improved morality in engineering are strangely silent about government agencies, stressing instead the conscience of the individual engineer in industry. They contend that as more and more engineers have become employees of large corporations, company loyalty has tended to overwhelm professional duty to the public. Where engineers do place loyalty to the public above loyalty to the company, they are defenseless against retaliation, usually in the form of dismissal. If engineers would be faithful to the public, so the argument goes, and if they could be protected against vengeful employers, many dangerous and wasteful technological products would never see the light of day. The issue is popularly known as "whistle blowing."

The idea of blowing the whistle on anyone, even when the public benefit is at stake, tends to make one uncomfortable. If an engineer can overcome this initial reluctance because he is determine professional obligation to society, the question of wl proper allegiance still remains thorny. Doctors and lawy uphold the public good, yet they have special oblig patients and clients. Surely engineers owe *some* special loyalty to their employers, particularly where patents and industrial secrets may be involved. However, assuming that proprieties could be agreed upon—a big assumption—engineers would still have to organize and bargain like labor unions in order to achieve the required protection from their employers. This has traditionally been considered to be unprofessional.

Some college professors have pointed to academic tenure as a model toward which engineers might strive. It is true that tenured professors with strong moral convictions are free to express their opinions without losing their jobs. This is fine for the professors. But it does not seem to have been very effective in solving society's problems. If the academic community, with all its freedom and moral stature, had made even the slightest progress toward providing a decent education for the masses, a good many of our crises, including environmental crises, could be speedily resolved. Moral convictions, and the freedom to express them, do not necessarily result in effective action, particularly when a lot of pontificating is going on.

But let us assume that some form of professional job security could be worked out, and that all engineers desirous of blowing whistles are concerned exclusively with the public welfare—not with their own aggrandizement or petty grievances. Fine. We would then be ready for action. We are agreed that we want to do the right thing. But how are we to determine what the right thing is? The thought of engineers designing and manufacturing weapons of war is horrifying to some of us, but many engineers consider such activity necessary and proper for the sake of freedom and national security. I would not like to be involved in marketing products that are not proven completely harmless under all conditions to individuals and to the environment; but many engineers consider it wrong to act the role of a Temperance Society, seeking to deprive people of products they desire, or even to drive prices sky-high by building in fail-safe features which people do not want. Certainly engineers in the automobile industry do not feel that they are betraying the public trust every time they design a car that has a lot of chrome trim, or is not as strong as a tank. It is possible, surely, to believe in good aith that people are entitled to have frivolous design, or that a totally safe car will encourage reckless driving, or even that making a high profit for a corporation is an important and acceptable thing to do.

We will search in vain for a single engineering moral absolute acceptable to the entire profession. There are a few cases upon which we might all agree—such as refusing to design gas chambers for a concentration camp—but precious few. Engineers are conservatives and radicals, hawks and doves, idealists and pragmatists—no need to extend the list in order to prove that engineers are people. Almost 15 percent of the profession are government employees; less than 5 percent are self-employed; approximately 7 percent are in education, or work for nonprofit organizations. The others—almost three-quarters of all engineers—are employed by private industry, and these range from presidents of giant corporations to frustrated young draftsmen, from scientifically oriented researchers and inventive geniuses to semiprofessional administrators and salesmen.[2] The multitude of conflicting objectives and beliefs of this diverse group can only be surmised.

We cannot expect to restrict the practice of engineering to those people whom we deem to have proper social and political attitudes. If we attempt to define engineering morality by our own philosophical prejudices, we will succeed only in fragmenting and politicizing our profession. Each of us must act according to his own conscience, of course. But we dare not begin accusing each other of lacking professional honor whenever we have a difference of opinion. We have seen people calling each other traitors because they have different views of how the nation should behave. By equating professional morality with our own political ideas, we are stooping to the same level. As engineers we are agreed that the public interest is very important; but it is folly to think that we can agree on what the public interest is.

We cannot even agree on the scientific facts! Would the SST adversely affect the upper atmosphere, allowing an excess of harmful ultraviolet rays to reach the earth? Will the large-scale use of nitrogen fertilizers or ordinary aerosol cans do the same thing? Will the Alaskan Pipeline do significant damage to the ecology of the wilderness? Has the total banning of DDT been precipitate, and done more harm than good? It depends upon which engineers you ask. There is scarcely a technical issue for which you cannot find expert witnesses of differing opinions. In questions of environmental impact, particularly, the complexities are so staggering that simple answers are unattainable. We are professionally committed to truth. But what *is* the truth?

Is the profession to establish a ruling body to decide? Are we to put our faith in an "academy"? Anyone who has read at all in the history of science and technology knows that the academy is wrong as often as it is right. Further, by its very nature as an established body, it tends to stifle originality and look after its own. The greatest minds of Britain's scientific establishment ridiculed Edison and said that his ideas for electric lighting of cities violated scientific possibilities. When Edison's reputation became established, he himself tried to have alternating current banned as unsafe. James Watt attempted to have the commercial use of high-pressure steam prohibited for the same reason. There are dozens of similar examples.

More recently, a committee of the National Academy of Sciences has bitterly criticized the research and development procedures of the Environmental Protection Agency. Scientists and statisticians employed by the Federal Food and Drug Administration have challenged the decision of their coprofessionals to approve the use of an artificial food coloring. An interdisciplinary team of scholars from Columbia, Harvard, and MIT has denounced as an "unwise gamble" antipollution regulations promulgated by federal experts, which were forcing the automobile companies into use of the catalytic converter. According to the National Bureau of Standards, at least 50% of the data reported in scientific journals are wrong or are supported by so little evidence that readers are unable to determine their reliability! The shining lights of our engineering and scientific academies are worthy of respectful hearing, but experience shows that their opinions cannot be taken as gospel.

Difficulties in assessing scientific and engineering facts are compounded by the problems of evaluating dollar costs and social consequences. Here the margins of error have been so enormous that one wonders why the public has not lost all confidence in the community of experts. In 1965 New York State voters approved a billion-dollar bond issue for, in the governor's words, "the elimination of water pollution in New York's lakes, streams, and rivers" within six years. Seven years later another 1.15 billion-dollar bond issue was approved for the same general purpose. After another three years had passed, it was estimated by a state spokesman that to meet Federal EPA standards by 1983 would require an additional expenditure "on the order of 18 billion dollars."[3]

Lately we have been hearing about "technology assessment," policy studies which seek to determine the fullest range of impact of any proposed technological change. An Office of Technology Assessment was established in 1972 to serve Congress. Other organizations, principally government agencies and the National Science Foundation, have sponsored such studies on subjects ranging from seeding hurricanes and artificially increasing snowfall, to geothermal energy, biological pesticides and biomedical technologies. Responsible, careful, multidisciplined studies, using the latest techniques of modeling and systems analysis, are surely deserving of our fullest support and respect. They may literally be mankind's last, best hope.

But even these reports must stand the test of public skepticism and political debate. The idea—expressed these days in the hortatory concluding passages of so many earnest books—that teams of experts can provide the information needed to allow the public to make "informed" choices, underestimates, in my opinion, the propensity of the wisest of men to make grievous miscalculations, and the inclination of ordinary people to make choices contrary to what appear to be their best interests. It also underestimates the problems inherent in *interpreting* scientific facts even when they are not the subject of dispute. For example, when scientists and engineers agree that radioactive emissions from a given number of power plants would increase the number of cancer deaths in the nation by 8.7 per year, what does this *mean*? Statistically, it means that the increased mortality risk to an individual is the equivalent of being a fraction of an ounce overweight, or smoking 0.03 cigarettes per year, or driving a car one mile per year, or adding a little salt to his food. Thus some people can say truthfully that the plants will cause cancer, while others, with equal sincerity, can say that the risks are negligible and well worth taking. But this is digressing. The point to be made here is that in talking about technology assessment we are talking about improved techniques, not about an improvement in engineering ethics.

We dare not lose sight of the fact that challenge and controversy are essential to a healthy profession, and even more to a healthy society. It is destructive, as well as wrong, to impute lack of conscience to everyone who sees things differently than we do.

The Kennecott Copper Corporation prepares to embark on a multimillion-dollar program to eliminate 86 percent of a smelter's sulphur dioxide emissions. This will meet the requirements set forth by engineers of the Air Conservation Committee of the State of Utah. Engineers from the Federal Environmental Protection Agency claim that the proposed technique is not the one that they would recommend, and besides, they consider a 95.2 percent reduction desirable. An engineer representing residents of the area appears at a hearing to warn of the health hazards involved. A Kennecott spokesman, backed up by his own engineering staff, calls the federal demand "untimely, unreasonable, uneconomic, unscientific, and unlawful."[4] He claims that the theoretical diffusion modeling upon which EPA calculations are based is not a proven or exact technique, and that the new equipment being recommended is untested and likely to prove unreliable. An agreement must be reached between the parties, or else the issue will be decided in the courts. Who is to say which of the many engineers involved is ethically superior to the others? The one criterion by which each of the engineers must be judged is integrity, and it is entirely possible—indeed probable—that this is a trait shared by them all in roughly equal proportions.

If the engineering profession could conceivably pull itself together and reach unanimous conclusions on issues of morality and truth, would the public interest then be served? Like the longshoreman's union, which occasionally orders its men not to load ships of one nation or another because of political issues, the engineers would have powers that might prove to be extremely dangerous to the commonwealth. It has been a continuing fear in the United States that the military forces might some day become politicized. The consequences of this to a fragile democracy are not pleasant to contemplate. But the potential power of the military is no more forbidding than that of a united and politicized engineering profession. The banding together of engineers into a cohesive group with a single moral code may be the dream of some; to others it is the stuff that Orwellian nightmares are made of.

Other considerations come to mind. Stressing the moral dimension of each engineering task might well serve to diminish the engineer's effectiveness and blunt his enthusiasm. An engineer designing a monorail car for a rapid transit system cannot become an expert in acoustics, urban planning, and the habits of woodland birds, and at the same time be a top-notch designer of monorails. Nor can he do his best work if he is excessively apprehensive and anxious about the ethical value of his every move. He must have confidence that other members of society are doing their jobs, including planning cities and protecting birds. Perhaps his interests as a citizen lie in different directions—supporting the local chamber music society, or working with retarded children. We cannot expect him to make his own environmental impact study relating to each assignment he receives. Naturally, one would hope that he will not approach his work as an automaton, blind to its aesthetic and moral implications. Of course, should he happen to discover that some aspect of his work will subject the public to a previously unsuspected danger, he will be honor bound to speak up. But, in general, the project—whether a monorail, a refrigerator, or a mechanical toy—should be subject to guidelines established by public agencies and monitored by engineers in the public employ. The role of *guardian*, while inherent in the role of professional engineer, clearly is different in many respects from the role of creator. A delicate balance must be struck between two valid yet contradictory propositions, one which holds that protection of the public is the responsibility of every engineer, the other that safety engineering and environmental engineering are specialties that should not be assigned in some vague way to the profession as a whole.

I wonder if those who seek salvation for society in the moral conversion of the engineer have considered the fact that natural selection of the marketplace works *against* the voluntary doer of good deeds. The company that undertakes pollution controls because of moral conviction will be at a competitive disadvantage in its industry. The buying public has shown very little inclination to reward "good" companies at the cost of paying higher prices. "Money, after all, is money," as Theodore Levitt, professor of business administration at Harvard, has written. "Its most compelling characteristics are scarcity and difficulty of acquisition. This imposes a discipline on its private use that lofty social consciousness seldom sways."[5] Thus engineers who pressure their employers to spend money for the sake of some public good may be contributing in a very real way to the welfare of a less worthy competitor. And individual engineers who refuse to work on a project that they consider to be contrary to the public interest will make way for less sensitive engineers. Voluntary good works are to be admired wherever they occur. But in the world of industry, compulsory good works, ordered by legislation, provide much better protection for society, and for men of conscience within industry as well. It is toward sensible and workable controls that we should all be exerting our efforts, rather than relying on a corporate virtue that is not likely to appear until the day of judgment. Government controls depend to some extent on morality, of course. But they stem more from common sense than they do from conscience. They are essentially a consequence of enlightened self-interest.

It is difficult not to be misunderstood here. Of course I want engineers to be moral. I want young engineers, in particular, to be idealistic and to resolve to serve the public interest as they see it. But the main trouble with engineers has not been their lack of morality. It has been their failure to recognize that life is complex. For a century they put their faith, somewhat unthinkingly, in "efficiency" and "progress." Now there is a danger that the same sort of mistake will be made with an abstraction called "social responsibility." My warning is simply: Beware of slogans.

Sanctimonious slogans have a way of lulling well-meaning people, and at the same time providing self-seekers with means to frustrate the very controls that are most needed. Take, for example, a report entitled, "The Engineer's Responsibility in Environmental Pollution Control," submitted in 1971 to the government's Council on Environmental Quality by the National Industrial Pollution Control Council. The report is an amorphous collection of noble generalities. It conjures up a vision of a crusading army of engineers, thousands abreast, marching in unison. The banner of this army is "cooperation." Its mission is to "coordinate," "unify," "interact," "centralize efforts," and "pool resources." Its weapons are "shared objectives," "common goals," "interdisciplinary concepts and techniques." The cloud of pieties serves, not to enlighten, but to obscure the real truth, which is that environmental pollution control can never be achieved by the worthy sentiments of industrial spokesmen, but only by government regulation. Charles F. Chandler and his fellow municipal engineers of a century ago recognized this. Somehow in the profession's embarrassment and guilt over the adverse effects of technology it has fallen back upon moralizing. Like schoolboys of an earlier age, engineers are lined up at the blackboard writing a hundred times, "I will not be bad again."

It is time for the engineering profession to grow up. Its problem is not lack of morality, but rather lack of maturity. It is not cynical to point out that pledges of virtue have never solved our problems and will not solve them now. In the real world, through conflict, debate, and struggles of the political process, issues must be decided. Engineers will participate in the struggles in many different roles, as befits their many different interests, prejudices, and skills. Since moral truths, and even technical truths, are so elusive, we have no choice but to rely on an adversary process. Actually this is what we are doing already, for this is the way things work in a democracy. The public interest must be protected mainly by engineers hired by the public. (And here I would include the increasing number of "public interest groups," as well as government agencies.) All engineers, of course, must subscribe to high standards of competence and integrity. There is no harm in repeating this again and again and again. But, by and large, engineers do subscribe to such high standards and always have.

It is to be hoped that in response to our present needs, more public-spirited engineers will enlist in public service. This is particularly important now, because the talented engineers who competed vigorously for public works jobs during the Depression years are mostly retired. In New York City, a senior engineering administrator has called for the creation of "a continuing elite cadre of career professionals," and has persuaded the Rockefeller Foundation to support a scholarship program toward that end.[6] Of course, a pouring forth of idealistic engineers will be of little avail if they are not hired by society. In 1974, when federal experts stated that at least 900 engineers should have been working on water-supply problems at the state level, the actual number employed was 300. This in spite of the fact that a 1969 survey had indicated that 25 million Americans were getting substandard drinking water from public systems, that as many as 8 million were getting water that federal experts called "potentially dangerous," and that outbreaks of disease attributed to drinking water were occurring practically every month. The other 600 engineers were available, surely. But state governments were not putting them to work. At the same time a proposed federal Safe Drinking Water Act was staggering through its fourth year of congressional debate. The more we prattle about morality, the more the world shows us how complicated things really are.

It is claimed that the professional societies should take a more active role in fulfilling the engineer's moral obligation to the public. But here, also, the matter is much more complex than it might seem. There is no easy solution to be found in calling upon the organizations to "do the right thing." The various professional societies have not been lacking in expression of civic concern. In 1889 the American Society of Civil Engineers established a Committee on Impurities in Domestic Water Supplies whose activities contributed to the establishment of the first U. S. Public Health Service Drinking Water Standards. One of the incentives that led President Theodore Roosevelt to convene a 1908 conference on conservation of natural resources was a resolution from the ASCE calling for the provision of national forest preserves. In 1922 the board of directors of this same society urged the federal government to undertake "a complete investigation of the cause, extent, and effect of pollution of waters by industries, that methods of mitigating such evils be investigated, and that existing legislation be reviewed in order to determine what if any legislation is required to cope with the situation." Even as the current environmental crisis was developing after 1950, the engineering societies showed concern and attempted to take action. (We will consider the effects of these actions a little farther on.)

But since the professional societies represent large numbers of engineers, and since there is often violent disagreement among the members concerning what is "good," the societies are subject to severe limitations on their actions.

Consider some of the problems raised at a seminar on the subject of technology assessment sponsored in late 1974 by the Engineers Joint Council. If professional societies are to undertake assessment studies, who should properly pay the cost? One society indicated its willingness to subsidize several studies at an expense of up to $10,000 each. Another society declared itself unwilling to use the dues of members for such a purpose; it operates on the assumption that studies should be made by volunteer effort. (Disappointment was expressed in not being able to involve representatives of the social sciences "who have a reluctance to work without pay!") Large-scale assessments, costing hundreds of thousands of dollars, cannot, of course, be undertaken by the societies, but must be funded by the government. Some members expressed concern that the assessments would appear self-serving, calling for technical solutions that would benefit the profession financially. Other members were most concerned that the studies might recommend solutions detrimental to their personal financial interests. Engineers in the aerospace industry, particularly, have suffered grievously because of changes in national policy. (Massive unemployment considerations aside, what is the "morality" of space exploration?) Should assessment studies, if funded, be undertaken by universities or private engineering firms, between whom there exists a little-publicized guerrilla warfare? Finally, apprehension was expressed by several engineers that "technology assessment" will inevitably lead to "technology arrestment."

Take another example of controversy. Six professional societies representing more than 400,000 engineers have subscribed to a Statement on Federal Energy Policy Planning which, while encouraging energy conservation, calls upon the government to "permit the most effective exploitation of those energy sources we already have" and to "create a climate in which new technologies can be developed, and existing technologies enhanced and strengthened."[7] At the same time, individual members of these societies have their own differing opinions concerning which course should be followed. One engineer writes an impassioned letter to a professional periodical exclaiming, "I say no, sir. We must concentrate all our efforts into finding new, nonpolluting energy sources . . . now."[8]

Several members of the American Society of Civil Engineers formed a Committee on Social and Environmental Concerns in Construction. Their worthy purpose was to be vigilant in such causes as minimizing defacement of the landscape and reducing noise levels during the course of construction projects. They soon found that there are many engineers who consider economy to be the dominating criterion in such matters, not only for the entrepreneur, but for the benefit of the consumer as well.

There are amongst engineers tempestuous disputes about requirements for local licensing. (Considerably fewer than half of America's graduate engineers have professional state licenses.) There are proclamations calling for civic involvement by the engineer. In 1919 the ASCE Committee of Development stated, "The engineering profession owes a duty to the public which it is believed can best be discharged by every engineer in the civic work of his community." But too often such activity has involved the engineer in "serving" as a volunteer the very community that eventually becomes his client. The incrimination of well-known publicly involved engineers in the Agnew scandal has had a traumatic effect on the profession.

So the role of the professional societies is seen to be most complex, and not susceptible to simple solutions in terms of improved social responsibility. Nevertheless, an increase of interest and activity by engineers in the public arena will be good for society, even if engineers turn out to be less consistent than the public might have thought. The engineering approach cannot, as we have seen, solve all social problems. However, it makes an extremely valuable contribution to the public discourse.

It is to be hoped that particular groups of engineers will continue to follow their special interests—unionizing, conservation, consumerism, radical causes or conservative, antigrowth or progrowth, antimilitary or promilitary. If some engineers get together to blow the whistle on large corporations, that is all right, too. The corporations can surely take care of themselves. In all of this, however, we must be on guard against any group which seeks recognition as spokesman for "the profession," and then seeks to impose its narrow definition of engineering ethics on us all. Debate within the profession is to be applauded, but it must not degenerate into petty vendettas or holy wars. This is always a danger when people start to argue about morality.

For those of us who are in search of a new philosophy of engineering, the new morality of "social responsibility" appears, upon inspection, to be pretty shallow stuff. In a way, it would be nice if it were otherwise. There would be satisfaction in saying, yes, we have sinned, and now, as an act of contrition, we will proceed to rescue the world. But if we have learned anything from our decline and fall, it is that the world is much too complicated to be rescued by platitudes.

As I have already suggested, the current crisis in engineering should be the occasion for calm reflection. Instead of indulging in an orgy of penitence, we should be thinking soberly about our failures and about the lessons to be learned from them.

The first thought that comes to mind is that if we have failed in our endeavors, perhaps we should retire from the field. Have we made so many mistakes, and damaged the environment so badly that technological enterprise should cease? Can technology simply be abandoned, the way fascism is abandoned when the people overthrow a totalitarian regime? Hardly anybody thinks so. The proposition is not even worth debating. It would be fun, perhaps, to recount in sordid detail the ways in which society would begin to disintegrate within hours after the technologists walked away from their jobs. But that would be proving a point that needs no proving. Distasteful as it may be to some people, it is clear that our survival, and the salvaging of our environment, are dependent upon more technology, not less. In *The Limits to Growth,* the Club of Rome's report which so distressed the public when it was presented at international meetings in the summer of 1971, it was noted that technology was a necessary factor in solving the critical problems inherent in exponential population growth. "Technological advance," states the report, "would be both necessary and welcome in the equilibrium state." Indeed, conservation itself is a "technique," as are forestry, horticulture, sanitary engineering, and other endeavors dear to the heart of environmentalists.

Since technology cannot be abandoned, the next logical step is to see what can be done to avoid repetition of technological mistakes that have been made in the past. Toward this end, let us consider the types of mistakes that engineers have made, and the reasons for them.

First, there are the mistakes that have been made by carelessness or error in calculation. Occasionally a decimal point is misplaced, the mistake is not picked up on review, and a structure collapses or a machine explodes. Human error. It happens rarely, and much less frequently now than in times past.

More often, failure results from lack of imagination. The Quebec Bridge collapsed while under construction in 1907 because large steel members under compression behaved differently than the smaller members that had been tested time and again. The Tacoma Narrows Bridge failed in 1940 because the dynamic effect of wind load was not taken into account. Although designed to withstand a static wind load of fifty pounds per square foot, the bridge was destroyed by harmonic oscillations resulting from a wind pressure of a mere five pounds per square foot. We do not have to be too concerned about bridge failures anymore. (In 1869 American bridges were failing at the rate of 25 or more annually!) But the problem of reasoning from small to large, and from static to dynamic, is symbolic of the difficulties we face in designing anything in a complex, interdependent, technological society. The Aswan Dam is an example. As a structure it is a success. But in its effect on the ecology of the Nile Basin—most of which could have been predicted—it is a failure.

Finally, there are the mistakes that result from pure and absolute ignorance. We use asbestos to fireproof steel, with no way of knowing that it is any more dangerous to health than cement or gypsum or a dozen other common materials. Years later, we find that the workers who handled it are developing cancer.

Human error, lack of imagination, and blind ignorance. The practice of engineering is in large measure a continuing struggle to avoid making mistakes for these reasons. No engineer would quarrel with the objective of being more cautious and farseeing than we have been before, of establishing ever stricter controls, testing as extensively as possible, and striving for perfection. It is beside the point to say that everyone makes mistakes, that there is no such thing as perfection, nothing ventured, nothing gained, and so forth. In such matters the engineer is willing to be held to the strictest standards of accountability. Of course, the elimination of error, depending as it does upon additional man-hours of study, has a direct relationship to cost. The public must be willing to pay such cost. Sometimes the expense is not warranted. We know that every once in a while an old water main will explode in one of our cities, but we do not want to spend the money to replace all old mains. Yet, on balance, it would seem that, knowing what we now know, both the engineer and the public would be willing to invest substantially in added protection, particularly where the environment is concerned.

Having thought a bit about the mistakes that engineers are liable to make, we find ourselves confronted with an unexpected phenomenon. It appears that even if engineering mistakes could have been eliminated entirely, the environmental crisis would have occurred just the same! The environmental crisis is upon us not because of any single error, but because of an accumulation of apparently error-free decisions. Nature is resilient enough to cope with a few oil spills, with the effluence from a few factories, no matter how foul, or even with an Aswan Dam. This resilience of nature, this capacity to absorb and to cleanse, has been assumed as a given parameter by engineering designers, and for each individual designer the assumption was valid. But pollutants have been emptied into the environment from thousands of sources, the designers of each source not knowing how many *other* sources were active or being planned. Each team of engineers worked on its own project without taking into account what many such projects might do to the environment over a long period of time. Nobody was looking at the total picture. Nobody was calculating the eventual cumulative effect of many new technological developments, compounded by explosive population growth. Nobody—or practically nobody—was planning for the future. Obviously this failure did not result from individual design error. Yet the end result has been much worse than the consequences of any particular engineering error one can think of. Something other than engineering failure is at issue here.

The engineering profession is not on trial. It is our democracy that is on trial. Politicians, anxious to retain power, pander to the lazy and selfish desires of the citizenry. Leaders dare not call for sacrifice except when crisis makes such a posture politically popular. Government traditionally has shown little interest in long-range planning (except for the military) and little concern for the preservation of natural resources. In this the government accurately reflected the indifference of the people. In the United States, the Environmental Protection Agency was set up in late 1970, only at the eleventh hour, and in an atmosphere of crisis. Nor has our society's lack of foresight been limited to the environment. There are crises of population, race, crime, education, economics, and many more, which were not anticipated, and for which there was no adequate planning.

If engineers had been asked to look into the problem of overall protection of our resources, they would have been happy to do so. Indeed, many of them were anxious to do so, and said as much. If they had been commissioned to prepare reports, good reports would have been prepared. In the absence of such requests and commissions, there is no rational reason for relating the crisis to engineering error. It has been said that such an attitude of "responsive engineering" is not worthy of our profession. This brings us back to morality, the implication being that in the absence of public support engineers themselves should have initiated action to head off the environmental crisis. The trouble with that argument is the fact that in the real world of real people, effective action on complicated problems can be taken only if someone will pay for it. Even the inventor who works nights in his basement eventually needs financing to develop his product. Everyone wants a cure for cancer, but we do not berate individual doctors for not working on this problem. We all know that an organized effort is required, and that such an effort is only possible if sponsored by society.

Where unique individuals have embarked on one-man crusades for the public good, their success has depended upon gaining public attention, and subsequently upon obtaining financial support. Dr. Albert

Schweitzer would not have been much help in the jungle without a hospital and medicine. Ralph Nader would have been very limited in his effectiveness if he had not been able to raise funds to pay for the operations of sizeable task forces. Also, Schweitzer and Nader, aside from being rare and unique individuals, started out by attacking problems that could be seen and understood. The problem of how we are overtaxing our natural resources is so complex that individual engineers could not have studied it meaningfully. Engineers are not poets or composers who can create in a small cabin in the woods. Most of the engineer's work today must receive large-scale support from society. Technology assessment projects entail costs running into hundreds of thousands of dollars. Preparation of environmental impact statements for the Trans-Alaska Pipeline are alleged to have cost about $8,000,000. It is estimated that the Climatic Implications of Atmospheric Pollution program sponsored by the Department of Transportation cost $9,000,000.

Considering that society did not provide such support and showed practically no interest in preserving the environment, the role of the engineer during the precrisis years was not one to be ashamed of. As early as 1952, the American Society of Mechanical Engineers appointed a full-time executive secretary to administer the activities of its Committee on Air-Pollution Controls. This committee organized an educational program for the society's annual meeting of that year, and according to the society magazine, *Mechanical Engineering*, "the intense interest shown in the many discussions indicates the importance that engineers attach to the subject of air pollution control." The committee established a semimonthly publication called *Smog News*, which quickly developed a coast-to-coast circulation amongst concerned engineers and public officials. In March 1955 the committee sponsored the First International Congress on Air Pollution, attended by representatives from the United States, The Netherlands, Germany, Portugal, Italy, France, and England.

A year earlier, Mark D. Hollis, Chief Engineer of the U. S. Public Health Service, addressed a convention of the American Society of Civil Engineers on the impact of rapidly changing technology on the field of sanitary engineering. A decade before Rachel Carson and Barry Commoner were to sound the alarm, Hollis expressed concern about the possible effects of new synthetic chemicals being spilled into our rivers and lakes, about toxic substances being introduced into the air (including the "unknown synergistic combinations of these"), about radioactive pollution, and new problems in food sanitation. He called attention to "a rate of technological change unprecedented in history," and stated that "the total amount of research now conducted in sanitary engineering is far short of the needs." The speech was abstracted at length in the May 1954 issue of *Civil Engineering.*

In early 1955, *Engineering News-Record,* the widely circulated, influential, industry-oriented McGraw-Hill publication, called upon President Eisenhower to launch a water pollution abatement program with a "grand plan," similar to those which had given birth to the nation's expansion in highways and schools. "Those close to the water pollution problem," the editorial stated, "have not been able to educate the whole nation." A forceful political impetus was required. It never came. By 1958 and 1959 this publication, among others, was running a series of editorials and articles pointing to the losing battle being fought against pollution of the environment. I do not want to overstate the case. The articles, editorials, newsletters, conferences, and the like, hardly constituted a crusade on the part of the engineering profession to save the environment. More could have been done. All right, more *should* have been done. But at least there is evidence that genuine efforts were made by the profession. The public, however, was not interested. Instead, starting in 1960, the nation decided to pour billions of dollars into space exploration, commissioning the engineer to put a man on the moon in ten years, which to everyone's amazement he did.

Should engineers have said that they did not want to put a man on the moon, but preferred to work on protecting the environment? A few of them did say such things, as individual citizens and in organized groups. But should they have spoken out thus *as a profession*? It would have been like the judiciary declining to enforce certain laws because they thought that the laws should be changed. Professionals have an obligation to lead, but they also have a duty to serve. Having been served, society then has no right to blame the professionals for its own shortsightedness. In its feckless way, society likes to claim that it has been misled by its experts. But one need not be an engineer to be concerned about the environment, any more than one need be a criminologist to be concerned about a crime wave, or an educator to be concerned about the fact that children can't read. The formidable grassroots fight against the pollution of Lake Superior by the Reserve

Mining Company of Silver Bay, Minnesota, has been led by two women: a secretary and a hair stylist.

In addition to being shortsighted, society has a disconcerting way of unexpectedly changing its standards of taste. Unfortunately, the engineer has not had much success in anticipating such changes. We have highways running all over creation because engineers never dreamed that one day people would decide that highways were ugly and unpleasantly noisy. Dams have sprung up like mushrooms at night because engineers never dreamed that one day people would decide that the preservation of beautiful canyons was as important as producing cheap power. Even the pollution of water and air has resulted in part from the lack of public concern about ugly sights and smells. The national mood for over a century has been dominated by a boosterism which flowed *from* the people *to* the engineer. Engineering in the United States started with the demand of farmers and ranchers for access to markets, and the concern for markets has remained paramount.

It is irrational to blame the engineer for things that were done at the behest of society, now that society has begun to change its way of thinking. Implicit in such blame, let me repeat, is the ultimate compliment, the assumption that engineers, as a group, can establish the goals and standards by which society lives. This is the reason that engineers are tempted to accept the blame. The situation has its amusing aspect. No sooner have engineers begun to give up their messianic notions than they find them being adopted by their critics!

The concept, when thought about seriously, is ludicrous. Engineers do not have the power to make major decisions for society. Also, as we have seen, they do not represent a single social philosophy. Even where there is general agreement on social philosophy, engineers are as perplexed as anybody else about the directions in which society should move and the ways in which priorities are to be established. Right now the masses of urban poor, desperate for jobs and decent homes, could not care less about preserving virgin wilderness. An occasional oil spill washing up on a bathing beach is of much less concern to them than an increase in the cost of keeping warm during the winter. They are even willing to accept noise and dirty air—up to a point—rather than see limited funds go elsewhere than to the daycare nurseries and other projects that they deem important. The same is true of the rural poor throughout the world, starving and in need of fertilizers and irrigation projects. Here we are not talking about good versus bad. We are confronted by a multitude of good objectives competing for limited resources, and present needs conflicting with obligations to the future. The big questions of what to do next are not technical, or only partly technical. They are primarily *political.*

Neither the concept of engineering error nor the concept of engineering immorality seems to be useful in explaining the origins of the environmental crisis. The engineer's role appears not to have been as dominant as society and some engineers have supposed.

A paradoxical thought occurs. Perhaps what society needs is not more morality from its engineers, but less. Perhaps engineers have been too honest and sincere, too naïve, too *innocent* to function effectively. The world is run by politicians and entrepreneurs who are subtle and devious, aware of the ambitions and fears that motivate the average man, but ignorant and expedient when it comes to taking action. If the engineer were a bit more savvy, he would be likely to have more say in the way things are handled, and this would probably be most beneficial to everyone. At least, when the engineer sees that a mistake has been made, he is quick to adjust and recommend new approaches. He has become quite skilled at analyzing the likely consequences of alternative courses of action. Although he is not a prophet, his knowledge and analytical methods are sorely missing in policy-making councils. It is constantly being proposed that engineering advisory groups assist these councils in their deliberations. This goal, however worthy, is too modest. If engineers could add a measure of sophistication to their other attributes, and then move away from their drafting tables to infiltrate society as leaders of corporations, universities, government agencies, and community groups, society's chances of coping with its problems would be markedly improved.

In a strangely convoluted way, something like this may be about to occur. For many years there has been talk about the need for engineers to get a liberal education. At most engineering schools this meant adding to the curriculum a few courses in English composition and the social sciences. Naturally this just gave the engineering student more practical information and did little to broaden him as a person. But with the coming of the environmental crisis, and the resulting outcry for a new engineering morality, new life has been breathed into the idea of liberal learning for the engineer. The development is an encouraging one, but not because of the false notion that study of the humanities will improve the engineer's character and make him a better citizen.

This view shows a remarkable ignorance about the engineering student, who is as decent a chap as there is around, and even more about the liberal arts, which are anything but a Sunday-school text. Let us hear Jacques Barzun, one of our leading defenders of the humanities, on this point:

> The humanities will not rout the world's evils and were never meant to cure individual troubles; they are entirely compatible with those evils and troubles. Nor are the humanities a substitute for medicine or psychiatry; they will not heal diseased minds or broken hearts, any more than they will foster political democracy or settle international disputes. All the evidence goes the other way. The so-called humanities have meaning chiefly because of the inhumanity of life; what they depict and discuss is strife and disaster. The *Iliad* is not about world peace; *King Lear* is not about a well-rounded man; *Madame Bovary* is not about the judicious employment of leisure time.[9]

Barzun might well have gone on to mention the novels of Stendhal and Balzac, where greed and ambition seethe on every page, and the novels of Dickens and Thackery, where pretense and vulgarity fill the stage. And then he might have mentioned the cultural heroes of our own era: the aesthetes and revolutionaries, the hedonists and iconoclasts. There is much in world literature about nobility and virtue, but there is a good deal more about desire and deceit, passion and despair. The time is past when a liberally educated gentleman was nurtured on works such as Plutarch's *Lives*, which extolled good citizenship. Today a student of the humanities is exposed to the feverish realities of life as well as to its exalted ideals.

The engineer already knows a lot about restraint and cooperation. He is logical, sober, and well-meaning, a very good citizen. I submit that study of the liberal arts will rob him of his innocence, stain his character, make him less "moral"—or, at least, less naïve. And this is exactly what the engineer needs. In all of his thinking, henceforth, he will have to take into account the imperfections and the absurdities of his fellow human beings.

Then, subtle and worldly-wise, he will sally forth, rising through the Establishment to positions of power never held by him before. At the same time, the born leaders among our youth, seeing that engineers are no longer docile technicians, and recognizing that technology is the key to survival for civilization, will turn to engineering as a profession, away from law, politics, and finance. Engineering will be "where the action is." Engineers will become leaders, leaders will become engineers, and the world will have a better chance of avoiding disaster.

An unlikely scenario? Perhaps. But much less unlikely than that which holds the engineer morally and technically responsible for the environmental crisis, and then sees a possible solution to the crisis in the improvement of his character.

Sir Hugh E. C. Beaver, addressing the First International Congress on Air Pollution in 1955, traced the seven-hundred-year-long campaign against air pollution in England. Complaint after complaint, committee after committee, report after report—all were ineffectual, as the centuries passed, and conditions grew progressively worse. Finally the London Smog of 1952, with its horrendous 4,000 deaths, set the scene for a new investigating committee, which was chaired by Sir Hugh. The committee's report was well received, said Beaver, and led to effective action, not because the report was exceptional in any way, but because the public was, at long last, receptive. The lesson to be learned, according to Beaver, is that "on public opinion, and on it alone, finally rests the issue." Or, as Aldous Huxley remarked to an august audience at a 1962 Conference on the Technological Order: "Evidently we have to have a great many tremendous kicks in the pants before we learn anything."

If engineers are distributed in positions of responsibility throughout society, public opinion is bound to be more informed and responsive to real need than it is at present. To the engineers in his audience Sir Hugh counseled "consistence and common sense" in all attempts to influence the common man. Consistence and common sense, competence and integrity—plus a dash of astute sophistication—this is what the times demand of engineers.

The plaintive call for a new engineering morality expresses a yearning to return to a time when engineers fancied themselves, in words which we have already quoted, "redeemers of mankind" and "priests of the new epoch." With the religion of Progress lying in ruins about us, we engineers will have to relinquish, once and for all, the dream of priesthood, and seek to define our lives in other terms.

PART 2

4. ANTITECHNOLOGY

If we are determined to look unblinkingly at the decline and fall of the engineering profession, we must not think that the environmental crisis, and the public uproar which it has provoked, are all that need concern us. An even more serious indictment of engineering is to be found in a curious new philosophy that has been gaining currency since the mid-1960s. It is the doctrine that holds technology to be *the root of all evil.* The proponents of this view are not satisfied to say that technologists have been careless, foolish, or immoral. They see the source of society's problems and men's miseries as lying in the concept of technology itself. If this viewpoint is credible, and if it takes hold in the minds and hearts of men, then our search for a positive philosophy of engineering is doomed to be abortive.

There are certain mechanical analogies that can be applied to human behavior—most aptly, the pendulum. When faith in technology began to diminish, one could have predicted a gathering of momentum that would carry opinion to some opposite extreme. This has proved to be the case. From the euphoric enthusiasm of the Golden Age of engineering there has been, among certain intellectuals, a swing toward a fierce abhorrence of technology. The philosophical movement which ensued we can only call, for want of a more graceful term, antitechnology. This movement has swept far past the legitimate concerns all intelligent people have for the environment, even far past the aesthetic concerns that trouble many artists, and past the emotional concerns that trouble many young people. It is a doctrine that sees technology as the devil incarnate. If technocracy was the extreme point on one swing of the pendulum, antitechnology is its equivalent on the next swing. Technocracy, however, was quickly recognized as an extravagant view, whereas antitechnology seems to have gained respectability and shows signs of establishing itself permanently in our perception of reality.

The founding father of the contemporary antitechnological movement is Jacques Ellul, a theological philosopher, whose book, *The Technological Society*, was published in France in 1954, and in the United States ten years later. When it appeared here, Thomas Merton, writing in *Commonweal*, called it "one of the most important books of this mid-century." In *Book Week* it was labeled "an essay that will likely rank among the most important, as well as tragic, of our time." In a short period it became a work very much talked about and quoted.

Ellul's thesis is that "technique" has run amok, has become a Frankenstein monster that cannot be controlled. By technique he means not just the use of machines, but all deliberate and rational behavior, all efficiency and organization. Man created technique in prehistoric times out of sheer necessity, but then the *bourgeoisie* developed it in order to make money, and the masses were converted because of their interest in comfort. Throughout most of history, Ellul contends, man controlled technique and was not too much occupied with it. But today it has escaped from man's control. The search for efficiency has become an end in itself, dominating man and destroying the quality of his life.

The second prominent figure to unfurl the banner of antitechnology was Lewis Mumford. His conversion was particularly significant since for many years he had been known and respected as the leading historian of technology. His massive *The Myth of the Machine* appeared in 1967 (Part I: *Technics and Human Development*) and in 1970 (Part II: *The Pentagon of Power*). Although he had spent a lifetime expounding the wonders of technology, he had found himself, after World War II, increasingly distressed by "the wholesale miscarriages of megatechnics." His anguished complaints and dire warnings about the course of technological society received a great deal of attention. Each volume in turn was given front-page coverage in *The New York Times Sunday Book Review*. On the first page of *Book World* a reviewer wrote, "Hereafter it will be difficult indeed to take seriously any discussion of our industrial ills which does not draw heavily upon this wise and mighty work." The reviewer was Theodore Roszak, who, as we shall see, was soon to take his place in the movement.

The next important convert was René Dubos, a respected research biologist and author. In *So Human an Animal*, published in 1968, Dubos started with the biologist's view that man is an animal whose basic nature was formed during the course of his evolution, both physical and social. This basic nature, molded in forests and fields, is not suited to life in a technological world. Man's ability to adapt to almost any environment has been his downfall, and little by little he has accommodated himself to the physical and psychic horrors of modern life. Man must choose a different path, said Dubos, or he is doomed. This concern for the individual, living human being was just what was needed to flesh out the abstract theories of Ellul and the historical analyses of Mumford. *So Human an Animal* was awarded the Pulitzer Prize, and quickly became an important feature of the antitechnology crusade.

In 1970 everybody was talking about Charles A. Reich's *The Greening of America*. After sections had appeared in *The New Yorker* magazine, creating a minor sensation, the book was published and quickly went through a dozen printings. In paperback it sold more than a million copies within a year. Reich, a law professor at Yale, spoke out on behalf of the youthful counterculture and its dedication to a liberating consciousness raising. Since Reich found technology to be the *bête noire* that inhibits our emotional development, he immediately took his place as a high priest of antitechnology.

Theodore Roszak's *Where the Wasteland Ends* appeared in 1972 and carried Reich's theme just a little further, into the realm of primitive spiritualism. Roszak, like Reich, is a college professor. Unlike *The Greening of America*, his work did not capture a mass audience. But it seemed to bring to a logical climax the antitechnological movement started by Ellul. As the reviewer in *Time* magazine said, "He has brilliantly summed up once and for all the New Arcadian criticism of what he calls 'postindustrial society.' " Leo Marx, commenting on the book in *The Saturday Review*, pointed to the inroads that antitechnology had made in our thinking: "Hostility to the secular intellect and a yearning for mystical transcendence now are powerful tendencies in America."

There have been many other contributors to the antitechnological movement, but I think that these five—Ellul, Mumford, Dubos, Reich, and Roszak—have been pivotal. They make an unlikely combination: a theological philosopher, a historian of technology, a biologist, and two academic apologists for the counterculture. However, they are united in their hatred and fear of technology, and surprisingly unanimous in their treatment of several key themes. With the exception of Dubos, who speaks in tones of sweet reason, they are all inclined to make extravagant statements, and, again with the exception of Dubos, they have been subjected to criticism and even ridicule. Yet certain of their basic ideas are receiving much serious attention. The thoughtful engineer has no choice but to give these ideas a hearing.

A primary characteristic of the antitechnologists is the way in which they refer to "technology" as a thing, or at least a force, as if it had an existence of its own. In this they take their cue from Ellul. "Technique has become autonomous," he writes. "It has fashioned an omnivorous world which obeys its own laws . . ." This is not just a figure of speech; it is a serious definition. Repeatedly Ellul emphasizes that "technique pursues its own course more and more independently of man." Mumford speaks in the same vein:

> Not merely does technology claim priority in human affairs: it places the demand for constant technological change above any considerations of its own efficiency, its own continuity, or even, ironically enough, its own capacity to survive.

Dubos refers to an uncontrolled force which he labels "undisciplined technology." He longs for "a human situation not subservient to the technological order," and concludes with the plea that we "avoid the technological takeover and make technology once more the servant of man instead of his master." Reich is given to personifying technology in such statements as: "Technology has its own reasons for removing things from the culture. . . ." ". . . technology will dictate to man. . . ." "Technology and the market have made our choices for us. . . ." He does not shrink from saying that affection, music, dance, work, and religion "have been ravished by an expanding technology," a technology which he fears is becoming "an unthinking monster." Roszak is no less dramatic. He fears "the treachery of technology," and warns that technology "threatens to murder the flora and fauna of whole oceans."

Making due allowance for poetic license, it is clear in these repeated personifications that technology is considered to have an existence separate and distinct from individual human beings. Indeed, technology is thought of as something that, unless fought against, can *do things to* human society, such things as "claim priority," "take over" and "dictate," even "ravish" and "murder." Dubos muses that although technology cannot theoretically escape from human control, society feels threatened by technological forces "just as it was threatened by the raiding Norsemen and Saracens ten centuries ago."

This way of thinking has spilled over into common usage, so we are not surprised to see an advertisement that begins, "Technology has trapped us. . . ," or an article in a news magazine that says, "Technology is seen as a dangerous ally."

Having established the view of technology as an evil force, the antitechnologists then proceed to depict the average citizen as a helpless slave, driven by this force to perform work he detests. This work, according to Ellul, is "an aimless, useless, and callous business, tied to a clock, an absurdity profoundly felt and resented by the worker. . . ." For best results the worker "must be rendered completely unconscious and mechanized in such a way that he cannot even dream of asserting himself." Asks Mumford:

> What significance can be attached to the current routine of the office, the factory, the laboratory, the school, or the university, founded as they so largely are on the sterile, life-inhibiting postulates of the power system? . . . What plethora of material goods can possibly atone for a waking life so humanly belittling, if not degrading, as the push-button tasks left to human performers actually are?

According to Dubos, our institutions are designed to make human beings "more efficient tools of industry and commerce." Reich could hardly agree more. He sees us all as "prisoners of the technological state, exploited by its economy, tied to its goals, regimented by its factories and offices, deprived of all those sides of life which find no functional utility in the industrial machine." Roszak notes with irony that even the technocrats are trapped into working to the point of nervous collapse.

In the bleak world-view of the antitechnologists, after the average person has been driven by evil forces to perform work he abhors, he is driven by forces no less malevolent to *consume* things he does not want. It is a central dogma of antitechnology that the consumer buys, not what he truly needs or desires, but rather those products which the technological society happens to spew forth.

Again it is Ellul who shows the way: "If man does not already have certain needs they must be created. The important concern is not the psychic and mental structure of the human being but the uninterrupted flow of any and all goods which invention allows the economy to produce." Mumford agrees that "the capitalist economy" has "sought to further industrial expansion by erecting the dogma of 'increasing wants' as an indispensable basis for further industrial progress." He speaks mournfully of the "expanding body of consumers, sedulously conditioned by advertising and 'education' to ask only for those mass products that can be profitably supplied."

Dubos lists among the "special needs" created by our society such things as "a different dress for every day at the office, a playroom in the cellar, and a huge lampshade in front of the picture window." Reich echoes the theme: "The machinery turns out what it pleases and forces people to buy." As Roszak sees it, our "habits of consumption" are controlled by the big corporations who, in turn, are the agents of "the suave technocracy."

At the same time that they identify an anonymous technology as the source of these evils (and apparently oblivious to any inconsistency) the antitechnologists place the blame on a particular group of individuals: the Establishment. Strong, selfish men, making use of technology for their own benefit, are supposed to be forcing the masses of men to work and consume under conditions that can only be described as subhuman. According to antitechnological doctrine, the leaders of the Establishment are assisted in their nefarious work by a staff of deputies: the technologists. Sometimes, by a strange alchemy, the technologists themselves become the Establishment—called technocrats—using their esoteric knowledge to dominate their bewildered fellow citizens. This state of affairs, say the antitechnologists, tends not only to debase the quality of life, but to perpetuate itself in a form of technocratic totalitarianism.

In the nineteenth century, according to Ellul, "the world was divided into two classes: those who created the economy and amassed its rewards, and those who submitted to it and produced its riches."

"Technique shapes an aristocratic society, which in turn implies aristocratic government. Democracy in such a society can only be a mere appearance." Nothing can be done, says Ellul, "to bridge the gap between the intellectual incapacity of the mob of specialized workers on the one hand and the monopoly of technical means by a technical elite on the other."

Mumford speaks of the "directors of the power complex—the military, bureaucratic, industrial, and scientific elite," who through publicity and prestige "are inflated to more than human dimensions in order better to maintain authority." The scientists, abandoning their ancient ideals of independence and disinterestedness, have become the new priesthood for the Pentagon of Power which dominates society, merely "keeping up a pretense of representative government and voluntary participation."

Dubos, a mild and humanistic commentator, by far the least rancorous of the five we are considering, is loath to accuse the Establishment of villainous intent. But he cannot resist making an attack on the technical elite:

> So far, we have followed technologists wherever their techniques have taken them, on murderous highways or toward the moon, under the threat of nuclear bombs or of supersonic booms. But this does not mean that we shall continue forever on this mindless and suicidal course.

"The nation," according to Reich, "has gradually become a rigid managerial hierarchy, with a small elite and a great mass of the disenfranchised. Democracy has rapidly lost ground as power is increasingly captured by giant managerial institutions and corporations, and decisions are made by experts, specialists, and professionals safely insulated from the feelings of the people."

Roszak turns to this theme again and again, each time in more anguished tones:

> The technocracy . . . is a citadel of expertise dominating the high ground of urban-industrial society, exercising control over a social system that is utterly beholden to technician and scientist for its survival and prosperity. . . . To be an expert or (as is more often the arrangement) to *own* the experts in the

> period of high industrialism is to possess the keys to the kingdom.
>
> . . . those who are not part of this expanding universe of expertise live on the margins of contemporary culture . . . they must watch these grandly consequential activities like spectators at an incomprehensible performance. . . .

Another atrocity of which technology is accused is cutting man off from the natural world in which he evolved. "Man was created," says Ellul, "to have room to move about in, to gaze into far distances, to live in rooms which, even when they were tiny, opened out on fields. See him now . . . in a twelve-by-twelve closet opening out on an anonymous world of city streets." Mumford avers that "a day such as millions spend in factories, in offices, on the highway, is a day empty of organic contents and human rewards." This could have disastrous results, as he sees it, since the human species came into being amidst the abundant variety of the natural world, and if contact with that natural world is not maintained, "then man himself will become . . . denatured, that is to say, dehumanized."

To Dubos, also, the decrease in contact with nature has been devastating:

> Modern man is anxious, even during peace and in the midst of economic affluence, because the technological world that constitutes his immediate environment, by separating him from the natural world under which he evolved, fails to satisfy certain of his unchangeable needs.

Reich mourns the disappearance from our lives of such vital experiences as "living in harmony with nature, on a farm, or by a sea, or near a lake or meadow, knowing, using, and returning the elements." Roszak is concerned that "the whole force of urban industrialism upon our tastes is to convince us that artificiality is not only inevitable, but better—perhaps finally to shut the real and original out of our awareness entirely."

This subject seems to cause the antitechnologists particular distress. Dubos and Reich compare modern man to a wild animal spending its life in a city zoo. The architecture of modern cities they find "blankly uniform" (Mumford), "lifeless and gleamingly sterile" (Roszak). As for the attempts of city dwellers to make some small contact with nature, Ellul scorns "a crowd of brainless conformists camping out," while Dubos refers to the "pathetic weekend in the country."

Not only weekend outings, but all of modern man's leisure activities, are subjected to the most critical scrutiny by the antitechnologists. Predictably, they view these activities with pity and contempt. Television, according to Ellul, is enjoyed because man seeks "a total obliviousness of himself and his problems, and the simultaneous fusion of his consciousness with an omnipresent technical diversion." Are spectator sports popular with the average citizen? Mumford tells us that they are "watched by thousands of overfed and underexercised spectators whose only way of taking active part in the game is to assault the umpire." As for such innocuous amusements as a pinball game and a jukebox, Mumford labels them "disreputable" because they do not "promote human welfare, in the fullest sense." Does the common man enjoy riding in his automobile? Dubos tells us it "represents our flight from the responsibility of developing creative associations with our environment."

Reich speaks of the "pathos" of an old-fashioned ice cream parlor where families amuse themselves "amid a sterile model of the past." He tells of a young couple who ski, play tennis, and sail, but only "think" they enjoy these pastimes, when they are really playing out roles copied from the mass media. Roszak is irritated by people "devouring hot dogs and swilling soft drinks" while waiting to see Old Faithful erupt. He observes with disgust that "their eyes vanished behind their cameras" and that one young boy said, "Disneyland is better."

These, then, are the main themes that run through the works of the antitechnologists:

(1) Technology is a "thing" or a force that has escaped from human control and is spoiling our lives.

(2) Technology forces man to do work that is tedious and degrading.

(3) Technology forces man to consume things that he does not really desire.

(4) Technology creates an elite class of technocrats, and so disenfranchises the masses.

(5) Technology cripples man by cutting him off from the natural world in which he evolved.

(6) Technology provides man with technical diversions which destroy his existential sense of his own being.

The antitechnologists repeatedly contrast our abysmal technocracy with three cultures that they consider preferable: the primitive tribe, the peasant community, and medieval society.

Ellul notes approvingly that primitive man "worked as little as possible and was content with a restricted consumption of goods. . . . The time given to the use of techniques was short, compared with the leisure time devoted to sleep, conversation, games, or, best of all, to meditation." Mumford declares that the agricultural work of neolithic times "brought the outer and inner life into harmony." Dubos reflects sadly on how the northwestern coastal Indians, who used to have lengthy periods of leisure, "do not find the time to carve and to paint now that they have accepted the efficient ways of technological civilization!" Reich believes that early man "built his life around the rhythms of the earth and his mental stability upon the constancies of nature," a theme echoed by Roszak who maintains that primitive tribes lived in close company with the earth, "striving to harmonize the things and thoughts of their own making with its non-human forces."

As for the peasant, Ellul tells us that he "interrupts his workday with innumerable pauses. He chooses his own tempo and rhythm. He converses and cracks jokes with every passerby." Mumford contends that "the poorest peasant . . . is foot-free and mobile," while Dubos feels that "our ancestors' lives were sustained by physical work and direct associations with human beings"—this being in contrast to "our absurd way of life." Reich is pleased to report that "the oldtime peasant had very real capacity for a non-material existence," and Roszak laments the eventual passing of "self-determining rural life."

Fond as they are of tribal and peasant life, the antitechnologists become positively euphoric over the Middle Ages. Medieval society, according to Ellul, "was vital, coherent, and unanimous," and opposed technical development with "the moral judgment which Christians passed on all human activities." Mumford rhapsodizes over "The Medieval Equilibrium" in which "a nice balance was established between the rural and the urban, between the organic and the mechanical, between the static and the dynamic components." Dubos is impressed by the fact that during the Middle Ages "Christianity acted as a great unifying force by giving the people of Europe a few common aspirations and social disciplines derived from the love and fear of God." According to Reich, in medieval times, "when a very different consciousness prevailed, neither technology nor the market was permitted to dominate other social values. . . ." Roszak finds in medieval alchemy one of the last precious examples of the "magical worldview," the disappearance of which he deplores.

Recognizing that we cannot return to earlier times, the antitechnologists nevertheless would have us attempt to recapture the satisfactions of these vanished cultures. In order to do this, what is required is nothing less than *a change in the nature of man.* The antitechnologists would probably argue that the change they seek is really a return to man's *true* nature. But a change from man's present nature is clearly their fondest hope.

Ellul is pessimistic, implying that intervention by God might be required. But the other four are more sanguine. According to Mumford, the growth of technology has "produced alterations in the human personality," and "modified . . . man's internal character." He feels that it is essential to reverse the process, and that it is still not too late to do so. To this end he proposes that we study closely the creature that is man. "We must understand the organics and physics of personality as we first understood the statics and mechanics of physical processes."

Dubos concurs, and calls for the development of a "science of humanity" so that we can have "a better knowledge of what human beings require biologically, what they desire culturally, and what they hope to become." With this "new kind of knowledge," we could then proceed to provide environments that would "encourage the expression of desirable human potentialities." Reich looks forward to "that 'change in human nature' which has been sought so long," and discerns the coming of a "new consciousness." Roszak also speaks of "reshaping the consciousness of people." All agree that a necessary first step toward the reshaping of human nature is a disavowal of the goals of technology, or rather an exorcism of technology from the soul of man.

Although the antitechnologists are concerned mainly with the harmful *effects* of technology, they reserve some of their disdain for the *activity* itself, and for the men who make it their life's work. As seen by the antitechnologists, engineers and scientists are half-men whose analysis and manipulation of the world deprives them of the emotional experiences that are the essence of the good life.

Ellul maintains that without the charts and motors that make them feel important, technologists "would find themselves poor, alone, naked, and stripped of all pretentions." Mumford calls them "adepts in abstract thinking though often babies in terms of well-salted human experience," and suggests that the scientific way of thinking represents a neurotic inability to face life as a whole. Dubos, being a scientist himself, is less harsh in his judgment of his peers. But Reich and Roszak really let loose on this topic. According to Reich, the technologist is alienated from his true self and his true needs. He is uptight, lonely, inauthentic, unable to receive or give out sensual vibrations. He is not a real man. He is "a smoothed-down man." Roszak maintains that the scientist and the technologist, in their objectivity, are guilty of "single vision," and "seeing with a dead man's eyes." Unless we can throw off the psychic style of technocracy, he concludes, we will continue to suffer "the death-in-life of alienation."

Under bombardment from such contemptuous judgments, the structure of our belief in engineering threatens to come tumbling down. The towers of arrogance were demolished by the environmental crisis. Now the foundation walls of confidence are being shattered by the philosophers of antitechnology.

Perhaps it is just as well. It is evident that something was wrong with the structure; it needed rebuilding. If critics have attacked with a vengeance, at least they have brought us down to the hard substratum upon which must be founded any new conception of the profession.

5. "I REFUTE IT THUS"

If we are to build a new philosophy of engineering, we must start with a rebuttal of antitechnology. Conceivably we could let the argument go unanswered, except to respond that technology is a necessary evil. But that would not be very satisfying. Besides, it would not give expression to what we know in our hearts; that technology is not evil except when falsely described by dyspeptic philosophers.

In the often-repeated story, Samuel Johnson and James Boswell stood talking about Berkeley's theory of the nonexistence of matter. Boswell observed that although he was satisfied that the theory was false, it was impossible to refute it. "I never shall forget," Boswell tells us, "the alacrity with which Johnson answered, striking his foot with mighty force against a large stone, till he rebounded from it—'I refute it *thus*.'"

The ideas of the antitechnologists arouse in me a mood of exasperation similar to Dr. Johnson's. Their ideas are so obviously false, and yet so persuasive and widely accepted, that I fear for the common sense of us all.

For a long time, many foolish things were said in praise of technology that should never have been said. Even now the salvation-through-technology doctrine has some adherents whose absurdities have helped to inspire the antitechnological movement. Also the growth as a serious discipline of the long-neglected history of technology has deposited layer upon layer of subtle thought upon what was once considered a fairly uncomplicated subject. Add to these absurdities and subtleties the malaise that is popularly assumed to prevail in our society, and you have the main ingredients of antitechnology.

The impulse to refute this doctrine with a Johnsonian kick is diminished by the fear of appearing simplistic. So much has been written about technology by so many profound thinkers that the nonprofessional cannot help but be intimidated. Unfortunately for those who would dispute them, the antitechnologists are masters of prose and intellectual finesse. To make things worse, they display an aesthetic and moral concern that makes the defender of technology appear to be something of a philistine. To make things worse yet, many defenders of technology are indeed philistines of the first order.

Yet the effort must be made. If the antitechnological argument is allowed to stand, the engineer is hard pressed to justify his existence. More important, the implications for society, should antitechnology prevail, are most disquieting. For, at the very core of antitechnology, hidden under a veneer of aesthetic sensibility and ethical concern, lies a yearning for a totalitarian society. But I am getting ahead of myself.

The first antitechnological dogma to be confronted is the treatment of technology as something that has escaped from human control. It is understandable that sometimes anxiety and frustration can make us feel this way. But sober thought reveals that technology is not an independent force, much less a thing, but merely one of the types of activities in which people engage. Furthermore, it is an activity in which people engage because they choose to do so. The choice may sometimes be foolish or unconsidered. The choice may be forced upon some members of society by others. But this is very different from the concept of technology *itself* misleading or enslaving the populace.

Philosopher Daniel Callahan has stated the case with calm clarity:

> At the very outset we have to do away with a false and misleading dualism, one which abstracts man on the one hand and technology on the other, as if the two were quite separate kinds of realities. I believe that there is no dualism inherent here. Man is by nature a technological animal; to be human is to be technological. If I am correct in that judgment, then there is no room for a dualism at all. Instead, we should

> recognize that when we speak of technology, this is another way of speaking about man himself in one of his manifestations.[1]

Although to me Callahan's statement makes irrefutable good sense, and Ellul's concept of technology as being a thing-in-itself makes absolutely no sense, I recognize that this does not put an end to the matter, any more than Samuel Johnson settled the question of the nature of reality by kicking a stone. There are many serious thinkers who, in attempting to define the relationship of technology to the general culture, persist in some form of the dualism that Callahan rejects.

It cannot be denied that, in the face of the excruciatingly complex problems with which we live, it seems ingenuous to say that men invent and manufacture things because they want to, or because others want them to and reward them accordingly. When men have engaged in technological activites, these activities appear to have had *consequences*, not only physical but also intellectual, psychological, and cultural. Thus, it can be argued, technology is *deterministic.* It causes other things to happen. Someone invents the automobile, for example, and it changes the way people think as well as the way they act. It changes their living patterns, their values, and their expectations in ways that were not anticipated when the automobile was first introduced. Some of the changes appear to be not only unanticipated but undesired. Nobody wanted traffic jams, accidents, and pollution. Therefore, technological advance seems to be independent of human direction. Observers of the social scene become so chagrined and frustrated by this turn of events—and its thousand equivalents—that they turn away from the old common-sense explanations, and become entranced by the demonology of the antitechnologists.

Once mysterious "technology" is invoked as a deterministic force, it becomes no longer intellectually respectable to say that our automobile culture has grown because people have always wanted to do the things that automobiles now enable them to do. Even an engineering educator hastens to assure us that such a belief is a sign of immaturity: "Until he becomes a manager, the young engineer interprets his work largely as a means to ends chosen by other people. A neat separability of ends and means is basic to this view of the relation of technology and values." A more mature view, according to this professor, will not assume that technology is merely the servant of men's purposes, but will face the question; "What if technology is a generator as well as an instrument of those purposes?"[2]

The "young engineer" might argue that technology did not create in people the desire to move quickly and independently from one place to another. Such a desire has existed within the human heart for a long time. Technologists, knowing of this desire, were, in a sense, "commissioned" to invent the automobile. Today it is clear that people enjoy the freedom of movement of which they had previously dreamed. True, they are unhappy about traffic jams, accidents, and pollution, but they recognize that these unhappy developments result from human decisions, not from technological imperatives. With remarkable stubbornness, and contrary to technological good sense, people persist in drinking and driving recklessly, refuse to take commuter trains, even where good and speedy ones exist, and resist joining together in car pools, which would reduce traffic by more than half.

Another unfortunate development that was not foreseen is the transformation of the automobile from a plaything to a necessity through the growth of suburbia. But how can this be blamed on technology, when a technological search for efficiency would dictate that people live in the cities where they work? Something other than technology is responsible for people wanting to live in a house on a grassy plot beyond walking distance to job, market, neighbor, and school. Not that wanting to live in the suburbs is necessarily a bad idea. The trouble with it is that when myriads of people set about doing the same thing the result is a "sprawl" of shopping centers and gasoline stations.

However much we deplore the growth of our automobile culture, clearly it has been created by people making choices, not by a runaway technology. Some people have come to despise the automobile, but at the present time they are very much in the minority. As more and more citizens become disgruntled with the problems arising out of mass ownership of the automobile, they are beginning to pass new laws controlling its use, and also to "commission" the technologists to devise different types of vehicles for individual and mass transport. Indeed, the technologists are already at work on their new assignment.

Of course, this is a superficial view of what has happened, but less superficial, I submit, than the antitechnological view, which sees a malignant technology "creating" choked and bloodied highways while the populace suffers in bewilderment.

There has been some attempt to find a middle ground in this dispute by resorting to the concept of a "soft" determinism. According to this view, technology provides new alternatives to society, and society then chooses which new path to follow. But even this misses the point, which is that a basic human impulse precedes and underlies each technological development. Very often this impulse, or desire, is directly responsible for the new invention. But even when this is not the case, even when the invention is not a response to any particular consumer demand, the impulse is alive and at the ready, sniffing about like a mouse in a maze, seeking its fulfillment. We may regret having some of these impulses. We certainly regret giving expression to some of them. But this hardly gives us the right to blame our misfortunes on a devil external to ourselves. We might, with equal lack of sense, blame wars on warfare.

Four decades ago, long before he became depressed by the atom bomb and other unfortunate technological developments, Lewis Mumford saw things more rationally than he and his fellow antitechnologists do today:

> Choice manifests itself in society in small increments and moment-to-moment decisions as well as in loud dramatic struggles; and he who does not see choice in the development of the machine merely betrays his incapacity to observe cumulative effects until they are bunched together so closely that they seem completely external and impersonal . . . technics . . . does not form an independent system, like the universe: it exists as an element in human culture and it promises well or ill as the social groups that exploit it promise well or ill. The machine itself makes no demands and holds out no promises: it is the human spirit that makes demands and keeps promises.[3]

In recent times, Mumford has seen fit to absolve the human spirit of any consequential sins. He and his fellow antitechnologists would have us believe that most of the unpleasant aspects of our life are caused, not by the human spirit, but by "technology"—a demon, a force, a thing-in-itself. This logical absurdity, which has been working its way into our popular consciousness, is the first antitechnological myth to be resisted.

In addition to confounding rational discourse, the demonology outlook of the antitechnologists discounts completely the integrity and intelligence of the ordinary person. Indeed, pity and disdain for the individual citizen is an essential feature of antitechnology. It is central to the next two dogmas, which hold that technology forces man to do work that is tedious and degrading, and then forces him to consume things that he does not really desire.

Is it ingenuous, again, to say that people work, not to feed some monstrous technological machine, but, as since time immemorial, to feed themselves? We all have ambivalent feelings toward work, engineers as well as antitechnologists. We try to avoid it, and yet we seem to require it for our emotional well-being. This dichotomy is as old as civilization. A few wealthy people are bored because they are not required to work, and a lot of ordinary people grumble because they have to work hard. Sociologists report that "job discontent is not currently high on the list of American social problems."[4] Or further: "It is clear that classically alienating jobs (such as on the assembly-line) that allow the worker no control over the conditions of work and that seriously affect his mental and physical functioning off the job probably comprise less than 2 percent of the jobs in America."[5] When the Gallup Poll asks the question, "Is your work satisfying?" the response is 80 percent to 90 percent affirmative. More sophisticated measures of job satisfaction, to be sure, uncover a great variety of complaints.[6]

The antitechnologists romanticize the work of earlier times in an attempt to make it seem more appealing than work in a technological age. But their idyllic descriptions of peasant life do not ring true. Agricultural work, for all its appeal to the intellectual in his armchair, is brutalizing in its demands. Factory and office work is not a bed of roses either. But given their choice, most people seem to prefer to escape from the drudgery of the farm. This fact fails to impress the antitechnologists, who prefer their sensibilities to the choices of real people. As engineers, we cannot expect to resolve an enigma that is inherent in the human condition. We merely do what we can to solve real problems. We are interested in, and are participating in, industrial experiments which seek to make the work experience more fulfilling. We will leave to others lamentations for arcadian days that never were.

As for the technological society forcing people to consume things that they do not want, how can we respond to this canard? Like the boy who said, "Look, the emperor has no clothes," one might observe that the consumers who buy cars and electric can openers could, if they chose, buy oboes and oil paints, sailboats and hiking boots, chess sets and Mozart records. Or, if they have no personal "increasing wants," in Mumford's phrase, could they not help purchase a kidney machine which would save their neighbor's life? If people are vulgar, foolish, and selfish in their choice of purchases, is it not the worst sort of copout to blame this on "the economy," "society," or "the suave technocracy?" Indeed, would not a man prefer being called vulgar to being told he has no will with which to make choices of his own?

Engineers devote more time and energy to creating hi-fi equipment and concert halls than we do to creating motorcycles. But if people want motorcycles, we are happy to provide them. And we do not like being called technocrats for our pains.

Which brings us to the next tenet of antitechnology, the belief that a technocratic elite is taking over control of society. Such a view at least avoids the logical absurdity of a demon technology compelling people to act against their own interests. It does not violate our common sense to be told that certain people are taking advantage of other poeple. But is it logical to claim that exploitation increases as a result of the growth of technology?

Upon reflection, this claim appears to be absolutely without foundation. When camel caravans traveled across the deserts, there were a few merchant entrepreneurs and many disenfranchised camel drivers. From earliest historical times, peasants have been abused and exploited by the nobility. Bankers, merchants, landowners, kings, and assorted plunderers have had it good at the expense of the masses in practically every large social group that has ever been (not just in certain groups like pyramid-building Egypt, as Mumford contends). Perhaps in small tribes there was less exploitation than that which developed in large and complex cultures, and surely technology played a role in that transition. But since the dim, distant time of that initial transition, it simply is not true that advances in technology have been helpful to the Establishment in increasing its power over the masses.

In fact, the evidence is all the other way. In technologically advanced societies there is more freedom for the average citizen than there was in earlier ages. There has been continuing apprehension that new technological achievements *might* make it possible for governments to tyrannize the citizenry with Big Brother techniques. But, in spite of all the newest electronic gadgetry, governments are scarcely able to prevent the antisocial actions of criminals, much less control every act of every citizen. Hijacking, technically ingenious robberies, computer-aided embezzlements, and the like, are evidence that the outlaw is able to turn technology to his own advantage, often more adroitly than the government. The FBI has admitted that young revolutionaries are almost impossible to find once they go "underground." The rebellious individual is more than holding his own.

The Establishment has potent propaganda techniques at its disposal, but this is more than offset by the increasingly free flow of information that the Establishment cannot control. And, as in the case of criminals, anti-Establishment movements have been quick to turn new techniques to their advantage. A generation ago it was the labor unions. More recent examples are the civil rights movement, the students' antiwar movement, and women's liberation. If members of the Establishment are indeed trying to persuade the masses to consume an oversupply of shoddy merchandise, then the consumer movement is a response that can be expected to grow, using advertising to combat advertisers, lobbyists to combat lobbies.

Exploitation continues to exist. That is a fact of life. But the antitechnologists are in error when they say that it has increased in extent or intensity because of technology. In spite of their extravagant statements, they cannot help but recognize that they are mistaken, statistically speaking, at least. The world was not "divided into two classes" starting in the nineteenth century, as Ellul contends. Reich is wrong when he says that "decisions are made by experts, specialists, and professionals safely insulated from the feelings of the people." (Witness changes in opinion, and then in legislation, concerning abortion, divorce, and pornography.) Those who were slaves are now free. Those who were disenfranchised can now vote. Rigid class structures are giving way to frenetic mobility. The barons and abbots and merchant princes who treated their fellow humans like animals, and convinced them that they would get their reward in heaven, would be incredulous to hear the antitechnologists theorize about how technology has brought about an increase in exploitation. We need only look at the underdeveloped nations of our present era to see that exploitation is not proportionate to technological advance. If anything, the proportion is inverse.

As for the role of technologists in the Establishment, it is ironic to hear ourselves called "high priests" and "technocratic elite" at the very time that we are complaining of a lack of prestige and power. Talk to any engineer or scientist, look into any professional journal, and you will learn quickly enough that the centers of power lie elsewhere. Technologists are needed, to be sure, just as scribes were needed at one time, or blacksmiths, or millers, or builders of fortresses. Some engineers have moved into positions of responsibility in industry and government, but their numbers are small compared to leaders trained in the law, accounting, and business. In any event, real power rests, not with the technologists, or with any special professional group for that matter, but with the wealthy, the clever, and the daring—and with their friends—just as it always has. How blind must one be not to see this obvious truth?

Nor do the technologists lord it over their fellows from a "citadel of expertise." That this myth persists is difficult to comprehend, but it appears to have a special place in the hearts of antitechnologists. John McDermott, in 1969, wrote a piece for *The New York Review of Books* entitled "Technology: The Opiate of the Intellectuals." It received quite a bit of attention at the time, and has since become a standard point of reference in the antitechnology literature. In agreement with the authors we have considered, McDermott asserted that "we now observe evidence of a growing separation between ruling and lower-class culture in America, a separation which is particularly enhanced by the rapid growth of technology." This is happening, according to McDermott, because "almost all of the public's business is carried on in specialized jargon," and "the new new language of social and technical organization is divorced from the general population."

This is persuasive rhetoric, but not in accordance with the facts. My teen-age sons read articles in *Scientific American* on computers, quasars, and laser beams with much readier comprehension than most of the pieces they are apt to find in *The New York Review of Books.* A typical plumber or gas station attendant, or even a bank teller who owns a secondhand car, knows more about society's technical systems and "the august mystery of science" (McDermott's phrase) than any dozen Establishment people such as bank presidents, political bosses, and Mafia godfathers. I pick up a book entitled *How Things Work*, intended for children of primary school age. It contains straightforward discussions of electricity and magnetism, internal combustion engines and rockets. With the help of simple diagrams, it explains the workings of carburetors, thermostats, transistors, and dozens of other devices. Where is all the mystery?

There are obscure specialties, to be sure, more than there have ever been. There was, for a time, much concern about a schism between the "two cultures," a phrase made famous by C. P. Snow in 1959. But Snow observed in 1963 that the divide seemed already to be closing, and with the growth of public concern about ecology and conservation, the general public is becoming more conversant with technological subjects rather than less so. (This can hardly be said, however, about the mysteries of economics, political science, and contemporary developments in the arts.) The average citizen, seated in front of his television set watching Walter Cronkite demonstrate the details of a flight to the moon, is not aware that he is being exploited by the mandarins of the technological elite. Nor are the technologists aware that they are the new mandarins. Both conditions exist mainly in the bizarre nightmares of the antitechnologists.

Next we must confront the charge that technology is cutting man off from his natural habitat, with catastrophic consequences. It is important to point out that if we are less in touch with nature than we were—and this can hardly be disputed—then the reason does not lie exclusively with technology. Technology could be used to put people in very close touch with nature, if that is what they want. Wealthy people could have comfortable abodes in the wilderness, could live among birds in the highest jungle treetops, or even commune with fish in the ocean depths. But they seem to prefer penthouse apartments in New York and villas on the crowded hills above Cannes. Poorer people could stay on their farms on the plains of Iowa, or in their small towns in the hills of New Hampshire, if they were willing to live the spare and simple life. But many of them seem to tire of the loneliness and the hard physical labor that goes with rusticity, and succumb to the allure of the cities.

It is Roszak's lament that "the malaise of a Chekhov play" has settled upon daily life. He ignores the fact that the famous Chekhov malaise stems in no small measure from living in the country. "Yes, old man," shouts Dr. Astrov at Uncle Vanya, "in the whole district there were only two decent, well-educated men: you and I. And in some ten years the common round of the trivial life here has swamped us, and has poisoned our life with its putrid vapours, and made us just as despicable as all the rest." There is tedium in the countryside, and sometimes squalor. No poet has sung the praises of Tobacco Road.

Nevertheless, I personally enjoy being in the countryside or in the woods, and so feel a certain sympathy for the antitechnologists' views on this subject. But I can see no evidence that frequent contact with nature is *essential* to human well-being, as the antitechnologists assert. Even if the human species owes much of its complexity to the diversity of the natural environment, why must man continue to commune with the landscapes in which he evolved? Millions of people, in ages past as well as present, have lived out their lives in city environs, with very little if any contact with "nature." Have they lived lives inherently inferior because of this? Who would be presumptuous enough to make such a statement?

The common domestic cat evolved in the wild, but a thousand generations of domesticity do not seem to have "denatured" it in the least. This is not the place to write the ode to my cat that should someday be written. Suffice it to say that although she never goes out of doors, she plays, hunts, loves, and eats with gusto, and relaxes with that sensuous peace that is uniquely feline. I submit that she is not more "alienated" than her wild sister who fights for survival in some distant wood.

The antitechnologists talk a lot about nature without clearly defining what they mean by the word. Does nature consist of farms, seashores, lakes, and meadows, to use Reich's list? Does not nature consist also of scorched deserts, fetid tropical forests, barren ice fields, ocean depths, and outer space—environments relentlessly hostile to human life? If farms and meadows are considered "natural" even though they have been made by men out of the stuff of the universe, what is "unnatural"? A stone wall and a farm cottage are still "good," I suppose, but a bridge and a dam become "bad," and a glass building façade becomes unnatural and dehumanizing, even though the glass has been made by man out of the sands of the earth.

Must one be in the wilds to be in touch with nature? Will not a garden in the back yard suffice? How about a collection of plants in the living room? Oriental artists have shown us how the beauty of all creation is implicit in a single blossom, or in the arrangement of a few stones. The assertion that men are emotionally crippled by being isolated from the wilds is, as I have said, unwarranted because of lack of evidence. But more than that, it does not take into account the multitude of ways in which "nature" can be experienced.

If pressed, the antitechnologists might grudgingly admit that the harm of being separated from nature can be mitigated if the separating medium is graceful and in harmony with natural principles, say like the Left Bank in Paris, or the Piazza Navone in Rome. But they point to the modern city as the epitome of everything that is mechanical and antihuman.

I will not here embark on a discussion of functionalism and modern architecture. But I will note in passing that there are millions of families who have lived happy years in nondescript high-rise apartments, and millions of people who have spent pleasant working lifetimes in the most modern office buildings. To claim that such passive environments are emotionally crippling is not to state a general truth, but rather to exhibit a personal phobia.

I have seen early-morning crowds pouring into a Park Avenue office building, into a spacious lobby, via a smooth-riding elevator to comfortable offices with thick carpets and dazzling window views. I have heard them chattering of personal concerns, a boyfriend who called, a child who scratched his knee, a movie seen, an aunt visiting from out of town. These people are no more dehumanized by their environment than are a group of native women doing their laundry on the bank of a river. I have seen them at their work, remarkably free to move about, exchange gossip, and gather at the water cooler. I have seen them promenading at lunch hour, looking into gaily decorated store windows, boys and girls eyeing each other, in time-honored fashion. I have shared in coffee breaks, drinking lukewarm coffee out of traditional cardboard and plastic cups with as much gusto as a peasant drinking his *vin ordinaire.*

I have also seen these office workers on a Monday morning comparing sunburns and trading tales about picnics, hikes, fishing trips, and various other sorties into the out-of-doors. The average person is not as isolated from nature as the antitechnologists would have us suppose. Ah, but this "pathetic weekend," as Dubos has told us, is not a true or meaningful relationship with nature.

There is a fussiness about the antitechnologists' abhorrence of the city, as if the drama of life could not unfold in anything but an idyllic setting. Saul Bellow, one of our leading novelists, has taken a more robust position. In his view, mankind is not about to be initimidated by anything as insignificant as a technological landscape:

A million years passed before my soul was let out into the

> technological world. That world was filled with ultra-intelligent machines, but the soul after all was a soul, and it had waited a million years for its turn and did not intend to be cheated of its birthright by a lot of mere gimmicks. It had come from the far reaches of the universe, and it was interested but not overawed by these inventions.[7]

The next target of the antitechnologists is Everyman at play. It is particularly important to antitechnology that popular hobbies and pastimes be discredited, for leisure is one of the benefits generally assumed to follow in the wake of technological advances. The theme of modern man at leisure spurs the antitechnologists to such heights of derision that we cannot help but question their seriousness of purpose. Perhaps they are merely expressing a satirical impulse, from the barbs of which no human activity is exempt. I have a fondness for grand opera, yet I have seen this sublime art burlesqued by a dozen comedians—and I have been amused. In *Gulliver's Travels* Swift has shown us how human beings appear ludicrous when viewed from an aloof perspective, and grotesque when viewed from close up. If the antitechnologists wrote in this timeless tradition, it would be querulous to object. But satire is clearly not their mode. In dead earnest, and with a purpose in mind, they are determined to show that the ordinary man at leisure is a contemptible sight, and that he has been reduced to his lowly state by the trickery of the technocratic society.

There are many popular pastimes contemptuously referred to as mass-cult activities—bowling, for example—that are not to my taste. But how can one draw sweeping conclusions from such a fact? A joyous, obviously exhilarating hour in a bowling alley is certainly not inferior in the scheme of things to a torpid, nonattentive hour listening to string quartets. Also, is bowling, or any of the other pastimes that the antitechnologists disdain, inferior to entertainments of earlier days, such as bear baiting, cock fighting, and public executions?

In their consideration of recreation activities, the antitechnologists refuse to take into account anything that an actual participant might feel. For even when the ordinary man considers himself happy—at a ball game or a vacation camp, watching television or listening to a jukebox, playing with a pinball machine or eating hot dogs—we are told that he is only being fooled into *thinking* that he is happy.

It is strategically convenient for the antitechnologists to discount the expressed feelings of the average citizen. It then follows that (1) those satisfactions which are attributed to technology are illusory, and (2) those dissatisfactions which are the fault of the individual can be blamed on technology, since the individual's choices are made under some form of hypnosis. It is a can't-lose proposition.

Under these ground rules, how can we argue the question of what constitutes the good life? If most people are fooled into desiring things they do not really desire, tricked into thinking they are free when they are really enslaved, mesmerized into feeling happy when true happiness forever eludes them, then clearly we are in a sorry state. But if the people themselves do not agree that their contentment is misery, what are we to conclude?

A character in a Gide novel remarks about the moment he first realized that "men feel what they imagine they feel. From that to thinking that they imagine they feel what they feel was a very short step!" Between feeling and imagining one feels, "what God could tell the difference?"

The antitechnologists fancy themselves to be the gods who can tell the difference. They charge technologists with having formed an elite class. But what is a little extra knowledge about machines compared to the godlike knowledge that they claim for themselves? Is it not clear that they consider themselves to be the elite of all elites?

They have complained that in the scientific worldview the scientist, seeking objectivity, cuts himself out of the picture, ignoring his own passions. The antitechnologists, however, in painting *their* picture of the true world, see nothing wrong with deleting the average man's passions. "As for the mass of urban workers," says Mumford, "they must have viewed their dismal lot, *if they were conscious at all,* with a feeling of galling disappointment." I have added the emphasis to the phrase which expresses so perfectly the antitechnologists' total scorn for anything that the average man might think, if indeed they credit him with thinking at all.

The idea that a man of the masses has no thoughts of his own, but is something on the order of a programmed machine, owes part of its popularity with the antitechnologists to the influential writings of Herbert Marcuse. In his *One-Dimensional Man,* Marcuse voices the disgust and frustration that are central to the antitechnological movement:

> If the individuals are satisfied to the point of happiness with the goods and services handed down to them by the administration, why should they insist on different institutions for a different production of different goods and services? And if the individuals are pre-conditioned so that the satisfying goods also include thoughts, feelings, aspirations, why should they wish to think, feel, and imagine for themselves?

It is legitimate, of course, to speculate on the extent to which people's lives are dominated by debasing illusions. Ibsen's *The Wild Duck* and Eugene O'Neill's *The Iceman Cometh* are two dramatic works that deal with the theme of how our lives are made tolerable by self-deceit, and with the problem of what happens when simple people are abruptly confronted with "truth." But the antitechnologists are not creative artists speculating about the eternal problems of being human. They are polemicists determined to prove that life today is worse than it used to be. At the very least, one would expect them to give weight to such evidence as is available. However, they avoid the discussion of facts, preferring to rely on such subjective impressions as the "blank, hollow, bitter faces," that Reich fancies he sees on the white-collar and blue-collar workers of America.

When real people are actually asked about their lives, "they believe that technology is both good and bad, and for most of them . . . the good outweighs the bad."[8] In medical studies assessing the adverse impact on health of changes in a person's life, it has been found that timeless events such as marriage, divorce, and death in the family are far more significant than anything having to do with the rule of technology in the world.[9] This is not to say that there is no dissatisfaction with life or disenchantment with technology; we will come to that in a moment. But the essentials of life do not seem to have undergone changes as sweeping as the antitechnologists maintain.

I leaf through *The Family of Man*, a book reproducing the photographic exhibition assembled in 1955 by Edward Steichen. I see 503 pictures from 68 countries, representing man in every cultural state from primitive to industrial. I see lovers embracing, mothers with infants, children at play, people eating, dancing, working, grieving, consoling. Everything really important seems eternally the same—in cities and in jungles, in slums and on farms. Carl Sandburg's prologue attempts to put it into words: "Alike and ever alike, we are on all continents in the need of love, food, clothing, work, speech, worship, sleep, games, dancing, fun. From tropics to arctics humanity lives with these needs so alike, so inexorably alike." A few moments spent studying these photos make the attitudes of the antitechnologists seem peevish and carping. These are real people with real faces that give the lie to the antitechnologists' snobbish generalizations.

Steichen and Sandburg are yea-sayers, a refreshing and necessary breed to have around. That does not mean that there is no place for Cassandras. The antitechnologists have every right to be gloomy, and have a bounden duty to express their doubts about the direction our lives are taking. But their persistent disregard of the average person's sentiments is a crucial weakness in their argument—particularly when they then ask us to consider the "real" satisfactions that they claim ordinary people experienced in other cultures of other times.

It is difficult not to be seduced by the antitechnologists' idyllic elegies for past cultures. We all are moved to reverie by talk of an arcadian golden age. But when we awaken from this reverie, we realize that the antitechnologists have diverted us with half-truths and distortions. We can see no reason why the gratifications experienced in earlier cultures (if, indeed, they *were* experienced) should be considered real or valid, while the expressed gratifications of people in our culture are to be discounted. We cannot agree that an earlier culture "in a long-range sense reflects the beliefs and values of the people in it" (Reich), while our culture does not. If we have been "sold" on automobiles and television sets, were not these earlier men "sold" on rain dances and promises of heaven? Or, if they created their cultures to fill their needs, have we not done the same? Man creates and is created in a never-ending, complex process. But the way the antitechnologists have twisted this to their purposes is simply intellectual deception.

Setting aside this technique of applying double standards, it is fair to go on to ask whether or not life was "better" in these earlier cultures than it is in our own. How is one to judge? The harmony which the antitechnologists see in primitive life, anthropologists find in only certain tribes. Others display the very anxiety and hostility that antitechnologists blame on technology—as why should they not, being almost totally vulnerable to every passing hazard of nature, beast, disease, and human enemy? As for the peasant, was he "foot-free," "sustained by physical work," with a "capacity for a non-material existence"? Did he crack jokes with every passerby? Or was he brutal and brutalized, materialistic and suspicious, stoning errant women and hiding gold in his mattress? And the Middle Ages, that dimly remembered time of "moral judgment," "equilibrium," and "common aspirations." Was it not also a time of pestilence, brigandage, and public tortures? "The chroniclers themselves," admits a noted admirer of the period, tell us "of covetousness, of cruelty, of cool calculation, of well-understood self-interest. . . ."[10] The callous brutality, the unrelievable pain, the ever-present threat of untimely death for oneself (and worse, for one's children) are the realities with which our ancestors lived, and of which the antitechnologists seem totally oblivious.

There are aspects of earlier cultures that seem appealing, and to which we can usefully look in structuring our own lives. But the antitechnologists have distorted the picture shamelessly, glorifying the earlier cultures and at the same time defaming ours. Then, with the air of protecting the higher values and the nobler pursuits, they blame the fancied deterioration in society on the role supposedly played by technology.

6.
A DANGEROUS DELUSION

It is not my intention to assert that, because we live longer and in greater physical comfort than our forebears, life today is better than it ever was. It is this sort of chamber of commerce banality that has driven so many intellectuals into the arms of the antitechnological movement. Nobody is satisfied that we are living in the best of all possible worlds.

Part of the problem is the same as it has always been. Men are imperfect, and nature is often unkind, so that unhappiness, uncertainty, and pain are perpetually present. From the beginning of recorded time, we find evidence of despair, melancholy, and ennui. We find also an abundance of greed, treachery, vulgarity, and stupidity. Absorbed as we are in our own problems, we tend to forget how replete history is with wars, feuds, floods, plagues, fires, massacres, tortures, slavery, the wasting of cities, and the destruction of libraries. As for ecology, over huge portions of the earth men have made pastures out of forests, and then deserts out of pastures. In every generation prophets, poets, and politicians have considered their contemporary situation uniquely distressing, and have looked about for something—or someone—to blame. The antitechnologists follow in this tradition, and, in the light of history, their condemnation of technology can be seen to be just about as valid as the Counter-Reformation's condemnation of witchcraft.

But it will not do to say *plus ça change plus c'est la même chose*, and let it go at that. We do have some problems that are unique in degree if not in kind, and in our society a vague, generalized discontent appears to be more widespread than it was just a generation ago. *Something* is wrong, but what?

I would hesitate to speak out on so formidable and complex a question if at least part of the answer did not seem self-evident. Contemporary man is not content because he *wants* more than he can ever have. The story of Faust is thought of as a romantic legend, but it embodies a profound truth. *Homo sapiens*, through the evolutionary process, has developed a unique combination of curiosity, creativity, and daring. These traits have been responsible for his success, while also confronting him with many serious problems. Although he has created stable societies that showed very little cultural evolution for long periods of time, once he is exposed to a new possibility, man cannot resist sampling it. He *will* taste new fruit, forbidden though it may be. He may taste fearfully and hesitantly, but he will taste. If the elders hold back, then youth will break away. If conservatives preach caution, then radicals will arise. The new attraction might be glass beads for Indian braves or spices for Renaissance princes; it might be the idea of heavenly salvation for marauding Vikings, or the concept of equality for Russian serfs; a new vision for the artist, or new knowledge for the seeker after knowledge.

Man learned early that changes in his way of life could have unforeseen and adverse consequences. The antitechnologists think it very significant that this has happened in the case of technological developments such as DDT. But it is self-evident that actions have consequences, some of which may be unforeseen and undesired. Bring flowers into the house, and your aunt may have an asthma attack. Invite your neighbor for dinner, and he may run away with your wife. Preach a religion of love, and you may start a revolution. New technologies are only a part of man's elemental impulse to experiment. This impulse does lead man to invent, but of equal importance—and hazard—it leads him to explore, to create new arts, new religions, new ideas.

Man has always been afraid of his urge to do more and know more. His earliest myths attest to this fear: Adam and Eve, the Tower of Babel, Prometheus, Pandora, Icarus. But he is constitutionally unable to restrain himself,

He has been unable to resist the temptation to have many children, and as a result the world is beginning to get crowded. (Although it would be a nice point in favor of technology to mention modern birth-control methods, the evidence seems to be that people have always been able to control the size of their families when they *wanted* to.) He has been unable to resist the temptation to try whatever he sees that appears interesting or amusing or labor-saving.

Our contemporary problem is distressingly obvious. We have too many people wanting too many things. This is not caused by technology; it is a consequence of the type of creature that man is. There are a few people holding back, like those who are willing to do without disposable bottles, a few people turning back, like the young men and women moving to the counterculture communes, and many people who have not gotten started because of crushing poverty and ignorance. But the vast majority of people in the world want to move forward, whatever the consequences. Not that they are lemmings. They are wary of revolution and anarchy. They are increasingly disturbed by crowding and pollution. Many of them recognize that "progress" is not necessarily taking them from worse to better. But whatever their caution and misgivings, they are pressing on with a determination that is awesome to behold.

The newspapers report that the Bulgarian government, bowing to consumer discontent, is attempting to provide more and better washing machines. This is not "technique" run wild, or "the suave technocracy" exploiting the people. *This is Bulgarians wanting washing machines.*

It is common knowledge that millions of underprivileged families want adequate food and housing. What is less commonly remarked is that after they have adequate food and housing they will want to be served at a fine restaurant and to have a weekend cottage by the sea. People want tickets to the Philharmonic and vacation trips abroad. They want fine china and silver dinner sets and handsome clothes. The illiterate want to learn how to read. Then they want education, and then more education, and then they want their sons and daughters to become doctors and lawyers. It is frightening to see so many millions of people wanting so much. It is almost like being present at the Oklahoma land rush, except that millions are involved instead of hundreds, and instead of land, the prize is everything that life has to offer.

Now, at last, we can see what it is that motivates the anti-technologists. It is fear. They are terrified by the scene unfolding before their eyes. They see hordes of college graduates in New Delhi serving with frustration as government clerks while wanting to be senators or leaders of industry. They see blacks rioting in the ghettos for a share in American bourgeois pleasures and for status as members of the professions and the business community. They see throngs of

Japanese students parading through the streets of Tokyo calling for Lord knows what. They see bricklayers demanding more money than professors, and getting it. They see firemen and teachers on strike—everyone seeking a share of whatever it is that is available. This situation has been developing ever since man emerged as a species, but it has accelerated alarmingly in our time.

The antitechnologists are frightened; they counsel halt and retreat. They tell the people that Satan (technology) is leading them astray, but the people have heard that story before. They will not stand still for vague promises of a psychic contentment that is to follow in the wake of voluntary temperance. Desperately the antitechnologists try to sell their vision of the ideal society, a sort of Viennese operetta scene, with the good and gentle populace dancing around the Maypole while the important personages (presumably including the antitechnologists) look on benevolently. But man has not come this far through the evolutionary furnace to settle for a bucolic idyll.

And why should he? If I enjoy an evening at the opera and a vacation trip to London, why should not others want the same? Some human desires can be labeled as vulgar and foolish, smacking of conspicuous consumption. Some are excessive, and doomed to lead to frustration and unhappiness. But most people are in search of the good life—not "the goods life" as Mumford puts it, although some goods are entailed—and most human desires are for good things in moderate amounts. The problem arises only when we put all of these moderate desires together and find that there is not enough of the good things to go around. Naturally the problem gets worse as the population grows and the desires become more extreme.

I agree with the antitechnologists that the situation is frightening. Perhaps there is some truth in the proposition that the common man would be "happier" if he did not have the urge to scramble upward to a higher station in life. But this is irrelevant because the common man does have the urge. The situation is summed up in the title of a World War I song: "How Ya Gonna Keep 'Em Down on the Farm After They've Seen Paree?" By now most of the world has seen some version of "Paree," and we have no choice but to live with the consequences.

Schopenhauer warned us a century ago about the *will* divided against itself. Will presses forward relentlessly in each living creature, heedless of its inevitable conflict with itself. Will, or life-force, or human nature—call it whatever you like—is what is at the root of our problems. Technology is merely one expression of this force. It is illogical to place the blame on technology. Why not blame the impulse to seek beauty, which we call art, or the impulse to seek truth, which we call philosophy, or the impulse to seek the ineffable "all," which we call religion? These are the sources of man's dreams and desires. These are the urges that drive man ever onward and refuse to let him rest. Man's technological skills may be responsible for the invention of the automobile, but he wants it and uses it because of his craving for new experiences, experiences of which he can conceive only because of his highly developed aesthetic sense and existential yearnings.

Further, if we are considering the source of man's discontent, let us remember that it is art, philosophy and religion that have made promises that cannot be kept. Technology's promises can be fulfilled. Visions of beauty, truth, and eternal bliss can only be mirages. Therefore, added to our real problems are the frustrations that must follow when we recognize that our dreams of Paradise can never be realized.

And let us not forget education. Every classroom is a factory for restless and questing human beings. Education widens horizons, raises expectations, and whets appetites, not only for intellectual and spiritual experience, but for a share of worldly goods. The idea of universal education did not originate with technology, yet the antitechnologists see no inconsistency in blaming technology for the consequences of this revolutionary concept.

Turning from abstract concepts to actual people, is it not fair to ask which men are most responsible for our present dilemma, technologists or artists—Newton and Edison, for example, or Socrates (who taught that knowledge is virtue), and Rousseau (who said that one man is as good as another)? The question may not be easy to answer, but it serves to remind us that poets and philosophers have been in the forefront of man's rush into the modern world. The antitechnologists accuse advertising and propaganda of being the handmaidens of technology, the evil sirens that persuade people to "want" things they do not really want, all for the sake of the megamachine. But advertising and propaganda, as everyone knows, are created by turncoat poets and renegade philosophers, certainly not typically by technologists.

It is wrong, of course, to blame art, philosophy, religion, and education while defending technology, for man is a single organic whole, and his technology has played a vital role in his evolution. I make the artificial separation only because the antitechnologists have done so first, and the desire is strong to counter them at every point. In their refusal to face up to the true facts of our contemporary condition, and to the real reasons for their panic, the antitechnologists have felt the need for a scapegoat. Their choice of technology has been arbitrary and unfounded.

One of the most convincing examples of the antitechnologists' true feelings is to be found in their attitude toward tourism. What could be more reasonable than the desire people have to travel and see parts of the world different from their own? Yet this reasonable desire results in crowds of tourists, and this irritates the antitechnologists to the point of frenzy. Roszak is the most virulent, calling the tourist trade "one of the great evils . . . one of the most destructive forms of pollution," and referring to "idiocies like pre-packaged tourism (the chance to make an international nuisance of oneself)." But each of the others displays a similar attitude of annoyance. Ellul scornfully remarks that man "becomes a cosmopolite and a citizen of the world, less . . . through his own will and ideals than through the mechanical fact of easy transport." "Why, indeed," asks Mumford dourly, "should any government subsidize jet-age travel when the net effect is to ruin every landscape and every historic site to which we bring our jet liners and motor coaches?" Dubos expresses his distaste by commenting that "the overwhelming majority of urban dwellers . . . identify leisure time with essentially aimless movement." Reich includes "vacation trips to Europe" among the extraneous features of our "affluent American way of life," and mocks the young couples who "manage to travel to some off-beat place each year" because they think that it is the fashionable thing to do.

Which of us has not been jostled by crowds in the Sistine Chapel or Westminster Abbey, or even at Stonehenge, and wished that he could be alone to savor these awesome places? But where the average traveler might be somewhat disconcerted, the antitechnologists react to these crowds as to a personal affront. Consider the tragedy which has befallen Mumford:

> Until now Delphi . . . presented one of the most wonderful landscapes in the world: a landscape whose profound religious atmosphere remained, though the temples are ruined and the

> religion itself has passed away. But, speaking for myself, I don't dare go back to Delphi. I know that it has already become a parking lot, and in a few years all that made it so precious will, if our present habits continue, disappear. . . .

And pity poor Roszak for the dreadful indignity he has had to endure:

> A personal anecdote: a sign of the times. Several years ago, on a cross-country trip, my family and I were foolish enough to visit Yellowstone National Park during the summer crush. . . . There was not an inch of solitude or even minimal privacy to be found anywhere during the two days we stayed before giving up and leaving. Never once were we out of earshot of chattering throngs and transistor radios or beyond the odor of automobile exhaust. . . . Ah, wilderness. . . .

How nice it would be for a select few of us if Delphi and Yellowstone could be set aside for our personal enjoyment, with the masses restricted to places such as Coney Island, more suited to their coarser sensibilities. But unfortunately our fellow citizens will not grant us such special privileges. Certain places that cannot possibly handle crowds, or that would be seriously damaged by crowds, are restricted to select scholars or other people of special qualifications. But, in general, those of us who would be tourists have little choice but to learn to put up with others who have similar desires and equal rights. Fortunately, although Delphi may be crowded, there are a hundred other ruined Greek temples that are not, and if Roszak would only leave his car in the Yellowstone parking lot and hike a mile or two into the woods, he can still find all the wilderness he could desire.

If the situation is not quite as appalling as the antitechnologists make it out to be, certainly the annoyance and fear they feel does have some basis in fact. But since the cause of the problem is not technology, which can be restrained, but the pressure of human desire, which cannot be restrained, it is difficult to know what to do except to continue to muddle along as best we can.

The antitechnologists do not see things so pragmatically. In their apocalyptic view, technology has brought us to the brink of disaster, and only an abrupt change of course can save us. Their deceptive statement of the problem is bad enough, but their proposed solution—a change in human nature—is much worse. The antitechnologists' desire to change human nature follows logically from their fear of the accelerating demands of the multitudes. One feels that they would be relieved to have frontal lobotomies performed on all the grasping, ambitious, foolish people who will not hearken to the antitechnological prophecies. Failing this, one wonders how they expect the change in human nature to occur.

If the first step is to be a scientific study of human nature, as Mumford and Dubos propose, what a strange scientific study that will be. Since the antitechnologists have decided in advance what human needs are, and have also agreed that the average man has a mistaken idea of what these needs are, one can scarcely imagine the sort of experiments they would devise.

With or without such a science, we know what the antitechnologists want for mankind. They want serenity and spiritual peace. But man wants something more. He may seek serenity when he does not have it. But as soon as he has it, he becomes restless and seeks some new adventure. No single way of life can satisfy him ultimately, least of all a return to the simple, rustic routines of earlier times.

The narrator of Dostoevsky's *Notes From Underground* speaks truly:

> . . . man is a frivolous and incongruous creature, and perhaps, like a chess-player, loves the process of the game, not the end of it. And who knows (there is no saying with certainty), perhaps the only goal on earth to which mankind is striving lies in this incessant process of attaining, in other words, in life itself. . . .

It is interesting that Mumford and Dubos both refer to this Dostoevsky story, but misinterpret it to support their own view. They attribute the antisocial behavior of "the sniveling hero" (they both use the identical phrase) to his dissatisfaction with a technological society. But the point is that Dostoevsky's character rejects not only the order and comfort of a technological society, but *every* attempt to define his needs and prescribe for his desires. It is improper to conclude from his scorn for the bourgeois life that he would welcome the neo-primitive utopias conjured up by the antitechnologists. He would ridicule their efforts to find a condition of lasting contentment for man.

> . . . man everywhere and at all times, whoever he may be, has preferred to act as he chose and not in the least as his reason and advantage dictated. And one may choose what is contrary to one's own interests, and sometimes one *positively ought* (that is my idea). One's own free unfettered choice, one's own caprice—however wild it may be, one's own fancy worked up at times to frenzy—is that very "most advantageous advantage" which we have overlooked, which comes under no classification and against which all systems and theories are continually being shattered to atoms.

If the antitechnologists hope to change human nature, they will not find technology standing in the way, but rather the caprice of which Dostoevsky speaks, the contrariness which makes fools of those who think that they have found the one right way.

The antitechnologists might protest that they stand with Dostoevsky not against him. Is not their literature filled with praise of freedom? Yes, but it is a strangely restricted freedom—the freedom to pray or sing or dance or weave, for example, but not to go stock-car racing or visit Disneyland or Las Vegas; the freedom to plant barley or corn, but not to use a bulldozer or buy a new electric hair dryer. It is the half-freedom, the false freedom, of the benevolent despot.

The antitechnologists are not preaching totalitarianism. They are good and gentle men, humanists at heart. But their cry for "something like a spontaneous religious conversion" (Mumford), "a common faith" (Dubos), "Consciousness III . . . an attempt to gain transcendence" (Reich), "the visionary commonwealth" (Roszak) is a cry for a new "movement," and each new mass movement carries within itself the seeds of a new totalitarianism. Despots arise when certain conditions exist: widespread disillusionment with the existing society, identification of a scapegoat, and the dissemination by glib prophets of new visions of salvation. These conditions will be fulfilled if antitechnology continues to grow in popularity. Technology, needless to say, has been selected to play the role of scapegoat.

It is not the prophets who become the despots, but despots arise who take advantage of the conditions created by the prophets. Just as dictators appear on the left and on the right, in the wake of religious fanaticism or antireligious fervor, so might one rise in the wake of despair stemming from antitechnology. Fear and disgust are what count, along with a scapegoat and pie-in-the-sky promises. If, practically speaking, despotism does not appear imminent, still the antitechnologists are lending their voices to the chorus of frustrated and frightened people who have lost faith in our institutions, and who are creating the conditions in which some sort of demagogue can rise to power.

The antitechnologists explicitly disavow any aggressive revolutionary activity. They announce the need for radical changes in our way of life, but for themselves and their followers they propose a quiet disengagement. Mumford calls, not for a revolt, but rather for "a steady withdrawal of interest, a slowing down of tempo, a stoppage of senseless routines and mindless acts." Dubos echoes this theme: "At heart, we often wish we had the courage to drop out and recapture our real selves. The impulse to withdraw from a way of life we know to be inhuman is probably so widespread that it will become a dominant social force in the future." To Reich, also, disengagement is the recommended procedure: "The plan, the program, the grand strategy, is this: resist the State, when you must; avoid it, when you can; but listen to music, dance, seek out nature, laugh, be happy, be beautiful. . . ." Roszak muses in the same vein: "As the old Gnosis comes back to mind in our time, once again people become—like most religious communitarians of the past—pacifist and anarchist. They disaffiliate, decentralize, cultivate nonviolent relationships, look after their own needs."

What if large numbers of good people *did* begin to withdraw as recommended by the antitechnologists? Such withdrawal, accompanied by cries of alarm, would only serve to make it easier for a demagogue to come to power. If a single tyrant does not come forward, then all the petty tyrants—the most selfish and aggressive ones—will speedily take over everything they can lay their hands on. It is foolhardy to assume that human nature will change spontaneously, to think that lions will lie down with lambs, when all the evidence of history as well as the experience of our daily lives tells us that it will not happen. In recent years we have repeatedly seen the settlements of the so-called flower children harassed by predatory gangs. Aggression is not forestalled by amiability.

Reich contends that the human nature which evolved in a world of scarcity can change now that we live in a world of abundance. But this is nonsense, since we do not live in a world of abundance, and we can never hope to live in a world which contains an abundance of all the things that people want. People are capable of generous thoughts and noble actions, and all men of good will are engaged in the effort to increase the amount of human virtue in the world. But this effort, if it is not to degenerate into fatuous piety, must include a recognition of the egoistic aggressiveness which exists in the scheme of things. How strange it is that the antitechnologists, who are enamored of nature, and who readily accept the behavior of leopards and vultures, are repelled by the idea that man, for all his angelic qualities, is self-seeking and competitive. It is not cynical to accept man as he is; it is prudent, yes, but reverent, as well.

To the everlasting credit of man, he recognized early the imperfection of his character, and set about finding ways of coping with it so as to enable himself to live harmoniously in large groups. For a long time spiritual and moral imperatives were declared by a succession of high-minded prophets. But then the philosophers of the Enlightenment, having observed that one man's faith always seemed to be another man's heresy, and that evangelism inevitably led to crusades, wars, and inquisitions, declared that formal religious and philosophical movements must be prevented from playing a dominant role in the organization of the state. What a unique and noble idea that was! The American founding fathers, sons of the Enlightenment, believed that certain basic rights should be accorded to each citizen, but after that the course of civilization was to be left to the vagaries of human impulse, as expressed in the bittersweet phrase, "the pursuit of happiness."

We have been attempting to muddle along, acknowledging that we are selfish and foolish, and proceeding by means of trial and error. We call ourselves pragmatists. Mistakes are made, of course. Also, tastes change, so that what seemed desirable to one generation appears disagreeable to the next. But our overriding concern has been to make sure that matters of taste do not become matters of dogma, for that is the way toward violent conflict and tyranny.

Trial and error, however, is exactly what the antitechnologists cannot abide. Roszak speaks for them all when he complains of "the great paradox of the technological mystique: its remarkable ability to grow strong by virtue of chronic failure . . . the sum total of failures has the effect of increasing our dependence on technical expertise." Agreed. Each new thing we do must be observed, maintained, and its consequences coped with. But this is true of all activities, not only technology. Educators are constantly coming up with new theories on how to teach children to read and multiply, and new plans such as the one proposing open admissions to state universities. When the results begin to come in, including some that are unforeseen and disappointing, there is a flurry of controversy and a whole flock of new theories from the experts. The same bewildering oscillation occurs in psychiatry, in law, in economics—in life. The experts play an important role in all of this, but they are far from having complete control. Usually the decisions to be made are social and economic, as well as technical, so that political processes come into play.

There is an increase in crime, and some citizens suggest that the streets be better lit at night. Technologists are called in and propose the use of high-intensity sodium lamps that shed much light with reasonable economy. Everyone is delighted with the results until it is discovered many months later that the lamps seem to be adversely affecting the health of trees. Should the lamps be replaced with others less bright and more wasteful of our resources? Should the trees be moved or left to struggle along? Where will the money come from to do whatever is to be done? Next time we should be more careful, we all agree, but next time the problem will be a different one. Do we want a more sophisticated traffic light system for a better flow of cars, or do we want to ban cars from the city? Do we want an enormous medical center or several smaller ones? Do we want factories and offices near our homes or far removed? Thousands of problems, with myriads of possible solutions—and each solution with many potential consequences, some obvious and some completely unknown. A demand is made for better planning, but the planners are as fallible as anybody else. Witness how quickly their theories change in the fields of housing, mental health, criminal justice, narcotics control, child rearing, just to mention a few. All we can do is do the best we can, plan where we can, agree where we can, and compromise where we must. Even if we had the omniscience to foresee the consequences of all our acts, we would founder on our inability to agree on what man is and what he wants, what he will be and what he will want.

But the antitechnologists will have none of this, as Dubos makes clear:

> We may hope eventually to develop techniques for predicting or recognizing early the objectionable consequences of social and technological innovation so as to minimize their effects, but this kind of piecemeal social engineering will be no substitute for a philosophy of the whole environment, formulated in the light of human aspirations and needs.

This has a wonderful ring to it, as do so many of the antitechnologists' grandiose pronouncements. Yet it is such dangerous advice. The passionate search for "a philosophy" is the very thing that is most likely to lead us down the dreadful path of dogma and totalitarianism.

Our blundering, pragmatic democracy may be doomed to fail. The increasing demands of the masses may overwhelm us, despite all our resilience and ingenuity. In such an event we will have no choice but to change. The Chinese have shown us that a different way of life is possible. However, we must not deceive ourselves into thinking that we can undergo such a change, or maintain such a society, without the most bloody upheavals and repressions.

We are all frightened and unsure of ourselves, in need of good counsel. But where we require clear thinking and courage, the antitechnologists offer us fantasies and despair. Where we need an increase in mutual respect, they exhibit hatred for the powerful and contempt for the weak. The times demand more citizen activism, but they tend to recommend an aloof disengagement. We surely could use a sense of humor, but they are in the grip of an unrelenting dolefulness. Nevertheless, the antitechnologists have managed to gain a reputation for kindly wisdom.

This reputation is not entirely undeserved, since they do have many inspiring and interesting things to say. Their sentiments about nature, work, art, spirituality, and many of the good things in life, are generally splendid and difficult to quarrel with. Their ecological concerns are praiseworthy, and their cries of alarm have served some useful purpose. In sum, the antitechnologists are good men, and they mean well.

But, frightened and dismayed by the unfolding of the human

drama in our time, yearning for simple solutions where there can be none, and refusing to acknowledge that the true source of our problems is nothing other than the irrepressible human will, they have deluded themselves with the doctrine of antitechnology. It is a hollow doctrine, the increasing popularity of which adds the dangers inherent in self-deception to all of the other dangers we already face.

PART 3

7. OF DULLARDS AND DEMIGODS

Let us suppose that our argument so far has merit. We have defended the engineer against charges that he is evil. Can we go further and defend him against charges that he is dull? The antitechnologists have characterized him as an uptight, inauthentic person who sees with a dead man's eyes. Except for its poetic excess, this is not too different from the generally accepted view. Although technology may not be destroying the world, it is surely destroying the image of the technologist as a sentient human being.

Even people who bear technology no grudge tend to think of the engineer as someone who is practical, analytical, and nonemotional. Such studies as have been made of the engineering personality seem to bear out the stereotype.[1] Engineers are bourgeois in their lifestyles and middlebrow in their tastes. Their interests are mostly limited to mechanical-technical matters on the one hand and the athletic-outdoor life on the other. "Constricted interests are apparent in their relative indifference to human relations, to psychology and the social sciences, to public affairs and social amelioration, to the fine arts and cultural subjects and even to those aspects of physical science which do not immediately relate to engineering."[2] The typical engineer avoids introspection and dislikes ambiguity. Although he gets along well enough socially, he would rather deal with things than with human beings.

If such generalizations seem presumptuous, listen to the conclusions of a research group which analyzed and compared five psychological studies of the engineering personality:

> These five studies yield a high consistency insofar as the character traits which engineers have in common are concerned. This is the more remarkable because these authors studied engineers in different fields and by different methods and techniques. It is therefore probable *that unlike many other occupations where it is impossible to demonstrate any consistent trend as far as personality traits are concerned, the engineering profession—with the exception of research, administrative and sales specialties—is composed of a homogeneous group of men with a fairly narrow range of temperamental variation.*[3]

The unpleasant truth is that today's engineers appear to be a drab lot. It is difficult to think of them as the heirs of the zealous, proud, often cultured, and occasionally eloquent engineers of the profession's Golden Age.

This may be, in part, a result of the profession's increasingly scientific orientation. Since World War II, and especially in the aftershock of Sputnik, engineering education has moved toward greater reliance upon theoretical science. This means that there is now very little time available for the student to do such physical things as surveying or playing with motors. The image of the engineer has changed from that of a ruddy-faced chap rushing about in high-laced boots, to a man in a white coat seated motionless at a computer terminal. But we cannot place too much blame on science, particularly when we see so many scientists who are cultured and concerned in ways that engineers are not.

Part of the problem is surely the stultifying influence of engineering schools. In too many of these institutions, the least bit of imagination, social concern or cultural interest is snuffed out under a crushing load of purely technical subjects. This situation appears to be improving, although a whole generation of engineers has already been disfigured.

At bottom, however, the problem does not seem to be so much in engineering itself as in the type of young person who is currently choosing engineering as a profession. We do not find Hegelian philosophers in the field as we did a century ago. Nor do we find "many sons of leading American families" such as, one historian of American engineering tells us, entered the profession in the late nineteenth century.[4] We find an increasing number of engineering recruits coming from working-class families. The typical engineering student is the serious, intelligent, unexciting young person whose profile we have just discussed. Engineering does not create this sort of person. It is this sort of person who is choosing engineering.

This is not necessarily a bad reflection on engineering. A great faith can attract mediocre apostles. During the Middle Ages many ordinary folk enlisted in the Church because it was powerful, respected, and provided a comfortable haven. Mighty spirits had built the Church, and many commonplace spirits followed in their wake. The same thing has happened in the trade union movement and in revolutionary movements throughout history.

It is undiscerning, however, to write off today's engineers as mediocre. Whatever their shortcomings, as described by psychologists and humanists, their positive attributes are not inconsiderable. By every available measurement, they are intelligent—more intelligent than the average college student, which means a lot more intelligent than the average person. They exhibit a high level of mental energy. And they are dedicated to their work.

Engineers choose their profession primarily because it promises "interesting work."[5] They do not expect great monetary reward or job security. Their objective is not prestige. They rank engineering below most other professions in this regard. They are, as a group, relatively free of arrogance and affectation.[6]

Intelligent, energetic, unassuming people who seek interesting work! Can this be mediocrity? The very least that one can say about such people is that they have enormous potential for growth. Thomas Mann tells us that he wanted the hero of his novel, *The Magic Mountain,* to be simpleminded, innocent, and curious. So he made him an engineer. But he is curious, Mann says "in a high sense of the word," and "the ordinary stuff of which he is made undergoes a heightening process that makes him capable of adventures in sensual, moral, intellectual spheres he would never have dreamed of . . ."[7] A "heightening process" is all that is needed to transform today's engineers from the dullards that they appear to be into the demigods that they seem capable of becoming.

Not only do engineers seek interesting work; they find it. The typical engineer derives great personal satisfaction from his daily activities.[8] There is something unique about this person that drives him to seek in engineering the fulfillment of an inner craving. And there is something unique about engineering that provides this fulfillment.

But when it comes to expressing the nature of the craving, or verbalizing about the quality of the fulfillment, the engineer draws back. In two volumes of essays, *Listen to Leaders in Engineering* and *The World of Engineering*, thirty prominent engineers have written about their profession, particularly addressing young people contemplating a career in engineering. What better place to pour forth their innermost feelings about their profession, their passions and their enthusiasms? But what we find instead is restraint and impassivity.

There is a gratification, gravely expressed, in being of service to society. Engineering "can provide a life of genuine satisfaction in many ways, especially through ministering in a practical manner to the needs and welfare of mankind." (Vannevar Bush). There are "rewards" when the engineer "feels that he has built something of lasting benefit to mankind . . ." (Rolf Eliassen).

The engineer experiences "the satisfying reward of knowing that he has contributed to the advancement of all mankind." (Wernher von Braun). There are "material and spiritual rewards that can come from contributing to the welfare of society." (Philip Sporn). But this satisfaction seems to stem less from inner feelings than from the sense of obligation fulfilled. The engineer has a "responsibility to help society" and a "duty" to employ his powers "for the use and convenience of man." (James R. Killian, Jr.). "His opportunity for productive contributions is almost without limit; his obligation to judge wisely and imaginatively is profound." (Newman A. Hall).

These engineers seem pleased that their work is important. It is good to be doing "interesting, impressive, and important things." (F. E. Terman). "There are many things that can attract the young engineer to a particular field of work, and among these is certainly a sense of importance." (John R. Pierce). They are gratified that their education helps them to comprehend the physical workings of the universe. There is "the deep satisfaction that stems from an understanding of the world in which we live." (George E. Holbrook). And they enjoy the neat way in which this physical world is subject to manipulation. "There is real satisfaction in carrying this chain of events through to its end and seeing the machine jump through the hoop." (Edward E. David, Jr.).

There are a few references to "creativity," but they are subdued. There is much talk about "challenge," but the meaning of the word is never explored. There is "satisfaction," and occasionally "fascination." In a few, rare instances, engineering is called "exciting." The more usual adjective is "interesting." We are told that engineers "like" their work. We are told that they are "responsible, and reliable, and orderly, and . . . conscious of human values. . . ." (William O. Baker). We are told that they have a "deep-rooted drive to produce tangible, useful results. . . ." (M. P. O'Brien). "Above all, engineers want to do things that are useful and will work." (Edward E. David, Jr.).

The impression one gets from these two volumes confirms the reproaches of the antitechnologists. These great engineers—these talented and productive individuals—write of their life's work with scarcely a glimmer of passion. They appear to be, as Reich has described them, "smoothed-down men."

How would I expect them to write? There is one small example that shows what they might have said. In the last essay of *Listen to Leaders in Engineering*, just as the weary reader is dragging listlessly through the final pages of murky platitudes, he comes across these two sentences by Jerome B. Wiesner:

> Technical and scientific work is usually fun. In fact, creative technical work provides much the same satisfaction that is obtained from painting, writing, and composing or performing music.

Fun! What a startling word. Engineering is fun, and similar to the creative arts in providing fulfillment. Just two sentences. A brief, brilliant light illuminating the night, and then darkness closes in again.

Maybe it is unfair to expect these august gentlemen to exhibit fervor in their writings. Their very importance in the profession is undoubtedly an inhibiting force. Yet engineers of less reputation are equally restrained when addressing an audience. Styles have changed, of course, and the ornate prose of earlier times is out of favor. A president of the American Society of Civil Engineers could say in 1868, "The true engineer loves and is devoted to his profession. He believes it to be 'the noblest of them all,' giving scope to higher and deeper thoughts. . . ."[9] Such expression today would seem overblown. So would the Hegel-inspired writings of John A. Roebling, who died in 1869, before he could complete his opus, "The Harmonies of Creation." But while the trend in our culture has been away from pretentiousness and sentimentality, it has been toward an intense exploration of inner feelings. Somehow the engineer has veered away from the sentimentality of the past without showing any inclination to search for new personal truths.

Of course, engineering is not the only profession that lacks poets within its ranks. The romance of soldiering has been imparted to us, not by soldiers, but by minstrels. The cowboy is no more glib than the engineer, but he has been glorified in song and verse. Even the great lovers of history are known to us, not because of anything they said, but because of what poets have said about them. Henry V never exclaimed, "Once more unto the breach . . ." He might have been a very dull fellow. The engineer may merely be waiting for his Shakespeare.

Pending the arrival of a poet laureate for engineering, it is disheartening to see the stereotype becoming more and more fixed in the public mind. It is especially dismaying to see engineers contributing to their own caricature. Too many of them contentedly take practicality to be their province, while granting to the humanities a priority in the realm of emotion. By agreeing to this division, they unwittingly participate in creating the image of themselves as emotional cripples.

The women's liberation movement has made women conscious of the fact that their self-image is molded by society in a subtle and devious manner. Toys and story books make little girls think of themselves as "feminine" in ways that are socially acceptable but often based on distortions and misconceptions. Similarly, engineers are fooled into thinking of themselves in terms that are not inherently valid. It was useful for society to make women regard themselves in certain ways, and there is reason to believe that the same thing has happened to engineers.

In Maurice Samuel's novel, *The Devil That Failed*, the protagonist awakens in a strange room to find that he has been turned into a clumsy giant. He is dismayed by his plight. Gradually, with the aid of some rudimentary knowledge about the pendulum and the laws of gravity, he discovers that his size has not changed, but that he has been captured by a band of malicious midgets. We engineers have come to accept the image of ourselves as oafs, emotionally speaking. Some humanist midgets have helped to create this image in order to make themseves feel important. But we have done nothing to contest this defamation. In fact, we are continually making things worse.

The July 1972 issue of *The Atlantic* contained an article by Theodore Roszak entitled, "Science: A Technocratic Trap." We have already discussed Roszak's antitechnological views, so we should not be surprised to learn that his article evoked negative response amongst engineers. A professor of chemical engineering wrote a letter of protest to the editor, which was published along with a reply from Roszak. From the professor's message:

> , . . there have always been people who could not bear the pain of thinking, or of planning ahead, who wanted to return to a simpler era, to the safety of their grandfather's knee, to the womb. And such people have always resented the way that rational, foresighted people got ahead of them, and so they have cut them down to size by attacking their very base, their rationality. The sad thing, today, is to see some modern intellectuals subscribing to such patterns of attack, and actually cutting off their own roots in rationality.

And from Roszak's reply:

> It is not primarily science I pit myself against in what I write. Rather, the wound I seek to heal is that of psychic alienation; the invidious segregation of humanity from the natural continuum, the divorce of visionary energy from intellect and action.

Here we have a perfect example of what has been happening. Roszak condemns technology for ignoring the irrational elements of life. An engineer responds by defending rationality and accuses Roszak of being jealous of the successes of rational people. Roszak replies that he is sorry to see people of intellect and action isolated from "visionary energy," and that he is merely trying to heal the wound of psychic alienation. The engineer has helped Roszak make his point! Roszak was not attacking rationality. He was attacking the technologist for ignoring irrationality. There is a big difference.

Why could not the engineer have replied along these lines: "See here, Roszak, what makes you think that you have any sort of prior claim on visionary energy? Have you ever stopped to consider that those of us who are engineers relate to the world in ways that are not purely cerebral? Perhaps we live in closer touch with 'the natural continuum' than most people do, including most philosophers. Our lives are full of emotional experiences about which you know nothing. So do not seek to heal our wounds of psychic alienation until you have proof that we have suffered them. Perhaps we can teach you and your hip friends something about the good life."

But the engineer did not answer in that way. Instead, he stressed the very analytical, antiemotional aspect of his profession that irritates the humanist. Engineers are proud of their profession, anxious to sing its praises. But they cannot seem to get beyond perfunctory and nonpersonal expressions of the satisfactions they derive from their work. Many of them are "turned on" by what they do. But they are unwilling or unable to reveal their inner emotions to an audience.

The fact that engineers are inarticulate does not signify that engineering does not evoke strong emotions. The fact that engineers lack poetic flair does not prove—not by any means—that engineering has no soul.

We may be floundering at present in a dark night of the spirit. Our Golden Age has passed. Efforts to rally around a new morality of social responsibility are seen to be ineffectual and not well thought through. The attack of the antitechnologists can be rebutted, but the victory—if victory it is—produces no sense of exhilaration. Our fellow engineers lack passion and our spokesmen are totally uninspired. But this cheerless mood cannot prevail.

For engineering is an expression of mankind's most basic impulses and most sublime aspirations. Somehow the engineer will find a philosophical platform from which he can once again view the world with zest. He need not choose between the roles of discredited messiah and plodding technician.

The time has come for us to end with defending ourselves against spurious criticism and to start in search of an engineering philosophy for our age.

8. TOWARD AN EXISTENTIAL PHILOSOPHY OF ENGINEERING

Out of the cafés of Paris, shortly after the end of World War II, came word of a new philosophy called existentialism. Actually, this new philosophy was not new, having been derived from the writings of Sören Kierkegaard, a mid-nineteenth-century Dane, and developed in Germany after the First World War by Karl Jaspers and Martin Heidegger. Also, as noted in the preface, existentialism stems from an irrationality that predates all formal philosophy. Indeed, it could be said that this new philosophy was not a coherent philosophy at all, since the various existentialists seemed to agree on precious little other than their disdain for objectivity and scientific truth. But for all its lack of novelty, cohesiveness, and system, the existential movement very quickly became the talk of the intellectual world, its popularity enhanced not a little by such flamboyant proponents as Jean-Paul Sartre and Albert Camus.

In the years after World War II, two moods existed side by side in Western culture: an optimism that focussed on the opportunity to rebuild a ravaged world, and a deep pessimism that viewed the concept of progress as a false and discredited myth. As we have said, the Cold War, the hydrogen bomb, and the growing environmental crisis caused the gloomy view to prevail. Technology and social democracy seemed to be gods that had failed. Security and comfort remained ever elusive. Worse, in the few places where they were achieved, they seemed to do little to assuage men's anxieties. The very climate of disillusionment and alienation that brought about the end of engineering's Golden Age, furthered the growth of existentialism.

The essence of the existential view is disenchantment with conventional creeds, a resolve to dispense with comfortable delusions, and an insistence on looking inward for new truths. In Sartre's words, "subjectivity must be the starting point." Yearning, passion, anguish—these are what count, these are the realities that will not be denied. All philosophical systems, all scientific theories, all social programs are as naught compared to what each of us feels in his heart, in his bones, or in his gut.

For Kierkegaard a pervasive despair, which reason could do nothing to allay, led to a belief in God. "In his failure, the believer finds his triumph." To Sartre and Camus, only "the absurd" can be seen at the heart of things. It is the very lack of transcendent meaning that inspires them to make the most of their humanity. But among all existentialist philosophers—believers and atheists alike—there is a shared scorn for the truths of science and the promises of technology.

Existentialism has been blended into a mix with other social phenomena of our era—the Beat and the Hip revolutions, radical movements in the arts and religion, a renewed interest in Eastern philosophies, and of course the swelling tide of antitechnology. The word *existential* has come out of the mix somewhat altered and muddled, but grown in importance. More and more frequently we find it being used in everyday speech whenever someone wishes to refer to the innermost essence of the human spirit, those ultimate depths of our being that we are forever trying to understand and to satisfy. Rarely heard a generation ago, today the word seems to be everywhere. Norman Mailer calls the Hip movement "an American existentialism." He terms his political proposals "existential politics." He publishes a collection of his writings and entitles it *Existential Errands*, because he seeks "that moment which proves deeper than any of our pretenses." Rollo May reports that "existentialist influence" is seeping into American psychology and psychiatry. A psychiatrist confirms that he is increasingly consulted by patients suffering from a condition he calls "existential vacuum." We are not surprised to open a newspaper and find a Greenwich Village group advertising: "Existential Encounter . . . striving on a feeling level to know yourself."

To the engineer in search of a philosophy, existentialism seems at once appealing and alien. Its appeal lies in its honesty, in its courageous facing up to the reality of the inner self, its insistence on starting with *where we are* and *what we feel* rather than with comfortable shibboleths. But the existential search for inner truth suggests a sloppy emotionalism that appears to conflict with the engineer's reliance upon logic and the scientific method.

Yet, what if existential searching were to reveal at the core of the human spirit a love for engineering? Or what if engineers, seeking the basis of the satisfactions they derive from their work, were to come upon the very soul-satisfying elixir that existentialists prize?

My proposition is that the nature of engineering has been misconceived. Analysis, rationality, materialism, and practical creativity do not preclude emotional fulfillment; they are pathways to such fulfillment. They do not "reduce" experience, as is so often claimed; they expand it. Engineering is superficial only to those who view it superficially. At the heart of engineering lies existential joy.

It has become a cliché that technology is an obstacle in the path of those who seek to find an existential sense of themselves. The antitechnologists have preached this message, as we have seen. So have priests, philosophers, and poets. The argument seems self-evident. If people are entranced with trinkets, how can they plumb the depths of their spirit? If they are absorbed in analyzing systems and designing physical objects, how can they remain open for an encounter with Being? If they seek comfort, how can they expect to find truth?

These ideas seem so obvious, and have been accepted for so long, that we do not think to ask where they came from. Existentialists are not much given to looking for the historical origins of their ideas. But it is understandable, surely, if an engineer who would ally himself with the existentialists, and finds himself excluded from the club, so to speak, attempts to seek out the historical basis of this discrimination.

It is from the Greeks of the Periclean Age that we have inherited the idea that thinking is better than doing, and that pure thought is superior to thought sullied with utilitarian objectives. Plato advocated "the use of the pure intelligence in the attainment of pure truth." He even rebuked the astronomer for observing the sky rather than thinking about it, for "that knowledge only which is of being and of the unseen can make the soul look upwards, and whether a man gapes at the heavens or blinks on the ground . . . his soul is looking downwards. . . ." As for the "handicraft arts," they were beneath the dignity of the citizen, fit only for slaves. Plutarch, writing about Archimedes, assures the reader that, "regarding the work of an engineer and every art that ministers to the needs of life as ignoble and vulgar, he devoted his earnest efforts only to those studies the subtlety and charm of which are not affected by the claims of necessity." Philosophy, according to the Greeks, would lead to truth, truth was the equivalent of virtue, and virtue was the source of happiness. Aristotle reasoned that since the gods are "above all other beings blessed and happy," and since they could not be imagined as engaging in "trivial and unworthy" activities like men, they must be philosophers:

> Now if you take away from a living being action, and still more production, what is left but contemplation? Therefore the activity of God, which surpasses all others in blessedness, must be contemplative; and of human activities, therefore, that which is most akin to this must be most of the nature of happiness.

It is not surprising to find philosophers recommending the study of philosophy as a way of life. We see here, very early in world history, a phenomenon that is not often enough remarked: a handful of writers dominating the thought of a culture. We will never know whether or not the average Greek agreed with the philosophers of his time. Plato and Aristotle have the last word because they wrote down what they said. There is a nasty, unfair epigram which says, "Those who can, do. Those who cannot, teach." It could more accurately be paraphrased, "Many of those who can, do. Some of those who don't, write, and tend to criticize doing." It has been a complaint of historians of technology that the written record of the ages tells us much about writers and their patrons, and practically nothing about anybody else. It is still true today that writers reach bigger audiences than "doers." And writers, by virtue of their occupation and training, tend to have an "anti-doing" bias. Be that as it may, the Greek fondness for idealistic philosophizing haunts our public discourse to this day.

The other main source of antimaterialism in our culture is the New Testament. Its admonitions have echoed all around us for generations:

> For we brought nothing into this world and it is certain that we can carry nothing out. And having food and raiment let us be therewith content.

Even food and clothing, we are told, are not worthy of our attention:

> Therefore I say unto you, Take no thought for your life, what ye shall eat or what ye shall drink; nor yet for your body what ye shall put on. . . .

The lesson is repeated again and again in resounding prose. It is foolish as well as profane to be concerned with material goods, since they do not endure. Fire, rust, and moth are ever at the ready to destroy our handiwork. It is prudent as well as pious, therefore, to concentrate on thoughts of eternity.

The relationship of Greek and Christian idealism to the contemporary quest for existential truth is not without ambiguity. Although the Greek search for the ineffable All is in tune with the modern temper, Plato's interest in mathematics and Aristotle's interest in biology smack of the very scientific abstraction that existentialism deplores. Christianity's assumption that man is intended to dominate the earth conflicts with current ecological concerns. Also, the Christian attitude toward work is somewhat ambivalent. Man is instructed not to labor toward material goals, yet the Benedictines teach that idleness is the enemy of the soul. Often, indeed, the so-called Protestant work ethic is blamed for our worst technological excesses.

In the broad view, however, it can be seen that the effect of our Greek and Christian heritage has been to convince us that materialism is a defect in human nature. We refer to our materialistic society with shame. We feel guilty because we are not more spiritual. We feel elevated when we read a poem or look at a sunset. We become depressed when we compare our workaday routines with the glorious things that we might be doing. And yet. . . .

What is it precisely that we might be doing? It is all very well for Aristotle to maintain that God is happy immersed in contemplation, because nothing else is worthy of Him. But we are human beings, not gods. Most of us are not constituted to become yogis. We take another look into the depths of our being, a clear, hard, earnest, passionate look. We recognize that we cannot survive on meditation, poems and sunsets. We are restless. We have an irresistible urge to dip our hands into the stuff of the earth and to do something with it.

If we seek understanding instead of guilt, we need not stand in awe of Platonic idealism. Our heritage does not begin and end with the Greeks of the Periclean Age and the disciples of the New Testament. Of equal importance in the development of human society is the legacy of those ancients to whom material activity was the essence of life.

In this connection our first impulse is to think of the glories of Imperial Rome. Never was engineering more important or successful than during that era. However, there is something crass about that mighty empire that reflects and amplifies our concerns about our own civilization. It is better, I think, to move further back in time.

Long before Plato and Aristotle thought their lofty thoughts, the Greek city-states were establishing their power. For several hundred years before the Parthenon was built, Greeks sailed throughout the Mediterranean, establishing trading posts and fighting battles. At home they farmed, built harbors and cities, and perhaps, most important of all for their eventual prominence, mined silver and other metals. Slaves were gathered from all over the known world, gradually relieving Greek citizens of the need to perform onerous tasks. Out of the sweat and blood of generations of mariners, merchants, farmers, warriors, and technologists, came the wealth and the leisure that made possible the flowering of great art. In view of all this conflict and effort, the idealistic preachings of Plato and Aristotle take on the aspect of irony.

The deeds of the builders of Greek greatness might be lost in the dark recesses of history except for the genius of a single poet, the incomparable Homer. Four centuries before the flowering of Athenian culture, Homer created the epic poems *Iliad* and *Odyssey*. In these works are recorded forever the dreams and ambitions of a coarse, primitive, yet supremely dynamic civilization. Homer has been read and reread by every generation, not only because of his poetic talent, but because the world he depicts appeals to elemental human aspirations. The Homeric tales are as vital a part of our heritage as are the works of Plato and Aristotle. Our existential roots reach deep into those thunderous myths.

We all know of the mighty deeds of the heroes of the Trojan War. Much has been written about the wrath of Achilles, the nobility of Hector, and the resourcefulness of Odysseus. We know how those great warriors prized valor and reputation. It is less often remarked that they also loved *things*, the objects of their world. They relished the look of things, the feel of things, and the making of things. They were romantics, undeniably. At the same time they were the ultimate materialists.

In the world of Homer, men took delight in palaces, ramparts, harbors, and causeways; doors, thresholds, keys, and hardware; sailing ships, hawsers, oars, and ropes. There are bathtubs, thrones, tables, and footstools; armor, arrows, chariots, and harnesses; bowls, cups, pitchers, and baskets; cloaks, coverlets, robes, and tunics; axes, adzes, augers, and dowels. On virtually every page there pours forth a cornucopia of manufactured objects. And Homer must tell us about the materials of which each object is made. A sword is bronze with nails of silver; a helmet has a horsehair crest; a bow is made from polished horn; a chariot has an iron axle, and golden wheels with brazen rims; a shield is bronze with seven layers of oxhide; a corselet is cobalt, gold, and tin; a palace is constructed of smoothed stone, with roof timbers of fir, a threshold of ash, doorposts of cypress; a shelter is made with timbers of pine and a roof of thatch. There is fleece, spun cloth, amber, ivory, silver, olive wood, alder, and poplar. We can practically see and feel it all. The tin is "pale-shining," the leather is "gleaming," the wool is "soft," the oxhide thongs are "dyed bright with purple."

Then Homer insists on telling us how each object is made. The manufacture of Achilles' shield by the god Hephaistos is familiar to most students of literature:

> He cast on the fire bronze which is weariless,
> and tin with it
> and valuable gold, and silver, and thereafter set forth
> upon its standard the great anvil, and gripped in one hand
> the ponderous hammer, while in the other he grasped
> the pincers.[1]

In more than 130 lines of poetry the fabrication and decoration of the fabulous shield is described. At other points in the story we are told in detail about the manufacture of Pandoras's bow, Hera's chariot,

Odysseus's bed, and Achilles' shelter. We are informed that a door is made with double panels, that the pins on a robe open and close easily, that a key knocks a bolt upward, that tempering in cold water "is the way steel is made strong." When Odysseus sets about building a raft with the assistance of Calypso, Homer is in no hurry to get on with the plot:

> She gave him a great ax that was fitting to his
>   palms and headed
> with bronze, with a double edge each way, and
>   fitted inside it
> a very beautiful handle of olive wood, well hafted;
> then she gave him a well-finished adze, and led
>   the way onward
> to the far end of the island where there were trees,
>   tall grown,
> alder and black poplar and fir that towered to the
>   heaven,
> but all gone dry long ago and dead, so they would
>   float lightly.
> But when she had shown him where the tall trees
>   grew, Kalypso,
> shining among divinities, went back to her own house
> while he turned to cutting his timbers and quickly
>   had his work finished.
> He threw down twenty in all, and trimmed them well
>   with his bronze ax,
> and planed them expertly, and trued them straight
>   to a chalkline.
> Kalypso, the shining goddess, at that time came
>   back, bringing him
> an auger, and he bored through them all and pinned
>   them together
> with dowels, and then with cords he lashed his
>   raft together.
> And as great as is the bottom of a broad cargo-carrying ship,
> when a man well skilled in carpentry fashions
>   it, such was
> the size of the broad raft made for himself by
>   Odysseus.

Next, setting up the deck boards and fitting them
to close uprights
he worked them on, and closed in the ends with
sweeping gunwales.
Then he fashioned the mast, with an upper deck
fitted to it,
and made in addition a steering oar by which to
direct her,
and fenced her in down the whole length with
wattles of osier
to keep the water out, and expended much timber
upon this.[2]

We are mesmerized and begin not to care about whether Odysseus will ever use the raft to get home. The making of the raft seems to be all that matters. Homer's epithets are additional evidence of his feeling for the objects he introduces. The ax handle is "well-hafted" and the adze is "well-finished." His admiration for man's handicraft never flags. A helmet is "well-fashioned," pincers are "well-wrought," and a chair is "well-made." A shield is "a thing of splendor," or "a lovely thing." Everything is "splendid"—a door, a pitcher, a chariot yoke. A harbor is "a wonder to look at." Even a simple mule wagon is "a fine thing." Textbooks remind us that the epithets sometimes serve a purely poetic purpose in filling out the meter. But this qualification does little to diminish the overall effect.

In the Homeric world, objects have reputations, like the shield of Nestor, "whose high fame goes up to the sky. . . ." They have histories, even ordinary household items. A silver workbasket is brought into the room and we are told that it "had been given by Alkandre, the wife of Polybus, who lived in Egyptian Thebes. . . ." Pandoros pulls out his bow, and we are told how it was manufactured from the horn of a wild goat he had killed while hunting years earlier. In the midst of a battle the action stops while Homer tells us about a piece of armor, how the wearer's father received it as a gift—when, where, and from whom.

The giving and receiving of gifts is tremendously important. Extravagant gifts can propitiate anger. They are a token of true friendship. They bring honor upon the donor and represent respect for the recipient. They are an integral part of that hospitality which is the mark of a civilized man. Although a noble man is generous with his material possessions, he is energetic in his pursuit of them. He strives to win prizes in peace and in war, and considers it a sign of honor to strip the armor from a fallen adversary.

The pleasure that individuals take in particular objects does not obscure the appreciation of technology for its contribution to the well-being of the entire community. Protective walls were essential to the security of Thebes, "since without bulwarks they could not have lived, for all their strength, in Thebes of the wide spaces." A high value is placed upon maritime technology:

> For the Cyclopes have no ships with cheeks of
> vermilion,
> nor have they builders of ships among them, who
> could have made them
> strong-benched vessels, and these if made could have
> run them sailings
> to all the various cities of men, in the way that people
> cross the sea by means of ships and visit each other,
> and they could have made this island a strong settlement
> for them.[3]

Although the *Iliad* and the *Odyssey* are concerned mainly with the affairs of noble warriors, Homer makes clear repeatedly his admiration for the craftsmen of his age. He refers to "an expert carpenter, who by Athene's inspiration is well-versed in all his craft's subtlety." The manufacture of a shield is credited to Tychios, "far the best of all workers in leather." We are told of a famous smith, "who understood how to make with his hands all intricate things, since above all Pallas Athene had loved him." There is an architect fitting roof-beams, a bowyer making a bow, a swineherd who has built a handsome pig-sty. Armor is wrought "carefully" or "with much toil." A doorway has been "expertly planed." Women are given credit for creating lovely cloths and fine garments.

The gods, just like humans, enjoy their possessions and their opulent homes. Hephaistos, the artisan god, is extravagantly praised in most of his many appearances. He is "far renowned" and "strong-handed," famous for "his craftsmanship and cunning." As for his attendants,

There is intelligence in their hearts, and there
is speech in them
and strength, and from the immortal gods they have
learned how to do things.

We are accustomed to poets attempting to transcend material objects by transforming them metaphorically. Homer has a way of reversing this, bringing action and ideas back to materiality by means of simile. Hector's heart "is forever weariless, like an ax blade driven by a man's strength through the timber. . . ." Two men grapple with each other "as when rafters lock, when a renowned architect has fitted them in the roof of a high house to keep out the force of the wind's spite." An eagle sent from Zeus, instead of being the ultimate in symbols of the ethereal, has wings "as big as is the build of the door to a towering chamber in the house of a rich man. . . ." The goddess Athene, shedding grace on Odysseus, is compared to a master craftsman overlaying gold on silver.

We emerge from the world of Homer drunk with the feel of metals, woods and fabrics, euphoric with the sense of objects designed, manufactured, used, given, admired, and savored. If this be materialism, then our ideas about materialism seem to be in need of revision.

It can be said that the Homeric Greeks were crass and greedy barbarians, not worthy of our admiration, or that we, without their redeeming robustness, have become crass and greedy ourselves. Certainly one can get that impression watching, for example, a mob of shoppers storming a counter during a department store sale. But then, to disprove the glib generalization, there will be the shopper of sensibility, knowledgeable and alert, not driven by the demons of conspicuous consumption, selecting items with an eye for quality, utility, and beauty. The word consumer has taken on an unpleasant connotation, as if products were wolfed down by rapacious beasts. Manufactured items can be, and are, contemplated, appreciated, and admired.

Enlightened admiration can very easily drift into something akin to reverence. The object before us appears to be so lovely, or so grand, or so cleverly conceived that it inspires us, as does a work of art. When this occurs, materialism becomes imbued with something that is erroneously thought to be its opposite—spirituality. This occurs occasionally in

Homer, although, in the main, his materialism is unabashedly pagan. The fusion of material and spirit is found fully developed in another great source work of our culture—the Old Testament.

In the Old Testament, as in the *Iliad* and *Odyssey*, we find ourselves in an ancient, barren landscape, where man-made objects are the subject of wonder and delight. There is fond mention of bowls, dishes, caldrons, baskets, and vessels of all sorts; beds, chairs, couches, tables, candlesticks, and lamps; tents, curtains, robes, and sandals; trumpets, cymbals and harps. There are weapons, chariots and armor, although not nearly as pervasively as in Homer. The tools are the same as in Homer: hammers, tongs, anvils, nails, saws, and axes; also rule, line, and compass. Buildings and cities were as important to the Hebrews as they were to the Greeks; the Bible is full of walls, towers, warehouses, palaces, and gates. The desert environment makes for a special attitude toward water; the references to fountains, pools, wells and gardens are wonderfully refreshing. Canals, irrigation ditches, and troughs for animals take on an aura of importance. The love of materials is there, as in Homer: silver, gold, ivory and precious stones; iron and bronze; cedar, fir, pine, and olive wood; stone, marble and brick. Methods of fabrication are not dwelt on as lovingly as in Homer, although the description of the building of Noah's ark is detailed and exact, and the specifications for building the tabernacle, and later for building and rebuilding the temple, are exquisitely precise, taking up several chapters at a stretch. An appetite for fine workmanship and beauty is evident in many passages. The tabernacle has pillars overlaid with gold and "ten curtains of fine twined linen, and blue, and purple, and scarlet: with cherubims of cunning work . . ." Solomon's temple and palace are made of carefully hewed stones, planed and carved cedar, and ornately decorated brass columns that dazzle the Queen of Sheba. The palace of Ahasuerus is described as containing "white, green, and blue hangings, fastened with cords of fine linen and purple to silver rings and pillars of marble." The beds are "of gold and silver, upon a pavement of red, and blue, and white, and black marble."

Although considerations of beauty and luxury are important, the practical importance of technology is not overlooked. Judah won many battles, but even with the help of the Lord could not prevail over the inhabitants of the valley "because they had chariots of iron." Uzziah "made in Jerusalem engines, invented by cunning men, to be on the towers and upon the bulwarks, to shoot arrows and great stones withal. And his name spread far abroad . . ." He also dug many wells to provide for his large herds of cattle. Hezekiah diverted sources of water in order to foil the invading Assyrians in time of war, and to further the prosperity of Jerusalem in time of peace. He built storehouses for corn and wine and oil, stalls for cattle, and shelters for poultry. The mighty Solomon added to his wealth and power by building a fleet of ships. It is clear throughout the Bible that prosperity must be based on a vigorous technology.

Although kings get much of the credit for technological achievement during their reign, the technologists themselves are given ample recognition. Early in Genesis we are told of the birth of Tubal-cain, "an instructor of every artificer in brass and iron." Historical sources indicate that the name Tubal-cain refers to the smiths of Tubol, a district in Asia Minor where the smelting of iron is thought to have originated. When Moses builds the tabernacle in the wilderness, he relies upon Bezaleel "to devise cunning works, to work in gold, and in silver, and in brass, and in cutting of stones, to set them, and in carving of timber, to work in all manner of workmanship." In building the temple, Solomon sends for a master technologist from Tyre:

> . . . skillful to work in gold, and in silver, in brass, in iron, in stone, and in timber, in purple, in blue, and in fine linen, and in crimson; also to grave any manner of graving, and to find out every device which shall be put to him. . . .

When the temple is to be repaired during the reign of Josiah, money is paid "to the doers of the work. . . . carpenters, and builders, and masons . . ."

In all of this we are reminded of the world of Homer. There are other similarities, such as the giving of lavish gifts, mostly in the form of tribute to monarchs. But the materialism of the Bible is not to be confused with the materialism of Homer. The biblical love of physical objects is inextricably intertwined with the love of a single, almighty God. The Greek gods are given credit for teaching technological skills to men. But their role is that of friendly adviser and protector. The awesome God of the Old Testament has instilled in men creative skills so that they can build His holy kingdom on earth. Bezaleel, the multitalented builder of the desert tabernacle, does not merely appear upon the scene. He is sent to Moses by the Lord:

> And I have filled him with the spirit of God, in wisdom, and in understanding, and in knowledge, and in all manner of workmanship. . . .

The engineering impulse comes to man as a gift from God. Material enterprise is not to be shunned; it is to be pursued energetically, but with the service of God always kept uppermost in mind. The most worthy work is, of course, the building of tabernacles and temples and the bringing of offerings to the Lord. But technological effort directed toward prosperity for society is also considered worthy, if the prosperous society is to be devoted to virtuous purposes. Moses made it clear to the Israelites that they were being given a land of abundance, "a land wherein thou shalt eat bread without scarceness. . . . a land whose stones are iron, and out of whose hills thou mayest dig brass," but only on condition that they continue to worship the Lord and to abide by his commandments. If they should ever forget this commitment and begin to think, "My power and the might of mine hand hath gotten me this wealth," then "I testify against you this day that ye shall surely perish."

This theme is sounded repeatedly throughout the Old Testament. Pride in one's own accomplishments, and neglect of God's commandments, result in disgrace and destruction. A desire to further the will of God results in affluence. Uzziah, at the height of his power, having rebuilt Jerusalem and fortified it well, becomes vain, argues with the priests of the temple, and is immediately stricken with leprosy. When Solomon prays to God simply for "wisdom and knowledge," to lead God's people in the path of righteousness, he is granted not only what he asks, but also "riches, and wealth, and honour, such as none of the kings have had that have been before. . . ."

In the later books of the Old Testament—particularly Job and Ecclesiastes—weariness and a touch of cynicism begin to creep in. But throughout most of this vast and varied work, technology and material abundance are closely identified with piety and the highest good. Man is ordained to work, keeping always in mind the sanctity of the sabbath. Man is given the earth to explore and to enjoy—to exploit, if you will—but only if he does so in a spirit of righteousness.

Clearly Homer and the Old Testament leave us with a feeling about the existential quality of technology very different from that conjured up by the idealism of Plato and the New Testament. Instead of being a gross and desensitizing activity, technical achievement is seen to be the very essence of the good life—joyous, fulfilling, and holy.

I recognize that in a previous chapter I took issue with the antitechnologists over their idyllic descriptions of early cultures. I do not wish to be guilty of the same willful blindness. Doubtless life in Homeric and Biblical times was harsh and brutally primitive. But the message of Homer and the writers of the Old Testament is not one of abject accommodation to the existing scheme of things. They show us mankind unresigned, struggling and creating, striving to make a new world, and filled with lusty enjoyment of material creativity.

It may be said that the relationship of this primitive materialism to modern engineering is tenuous at best. Engineering today is based upon scientific principles which were undreamed of three thousand years ago. Also the contemporary engineer deals not only with metals and fabrics and building materials, but with such intangibles as gases and electrons—even with abstractions called systems and programs. But I believe that there are talents and impulses deep within us that underlie all of our technological creations—from towers and sailing ships to transistors and systems analysis. Primitive technology is very much the father of modern engineering. The main goal has always been to understand the stuff of the universe, to consider problems based on human needs (or desires), to propose solutions, to test and select the best solution, and to follow through to a finished product. Existential delight has been the reward every step of the way—for the observer, the user, and particularly for the doer. If a contemporary engineer feels a kinship for the technologists of the ancient world, who is to deny his right to assert it? If he says that he experiences an existential thrill in designing, fabricating and using physical objects—or even in solving a problem using a computer—then we must believe him. We cannot prove him right or wrong by taking his pulse!

Obviously, our forebears of Homeric and Biblical times experienced this thrill, and the capacity to experience it is still within us. It can be argued that we are so sated with inventions and gadgets that the ancient delight is no longer available to us. But if we are still able to delight in a birdsong, or a glass of wine, or a smiling child, why should we not still be able to delight in a well-made object or in an elegant solution to a problem? We are not as naïve as we were, but we have not been stripped of all enthusiasm and capacity for wonder. Poets are continually trying to revive our waning sensibilities. They could well take a lesson from the ancients, and refresh their own sensitivity to man-made things.

It can be argued that mass production has put an end to the fine craftsmanship that so pleased the Greeks and Hebrews. This is true, although if craftsmanship is not much in evidence, we can still enjoy the features of good design and superior quality.

It will be claimed that the ancients were able to take delight in their fabricated objects because they were not baffled by them. The work of the carpenter, the weaver, and the smith can readily be seen and understood. There is little mystery in the technology of chariots and armor. The obvious answer to this is that people today would get more pleasure out of the world if they understood more about science and technology. A good education should include enough in these areas so that the ordinary citizen is not deprived of his birthright, which includes savoring the engineering creations of his world. For today's engineer, as for the artisan of ancient times, the special knowledge and skill of the professional provide an added measure of existential delight.

A rereading of Homer and the Old Testament is the first step toward the formulation of an existential philosophy of engineering. These works reinforce our intuitive belief that engineering is a basic instinct in man, the expression of which is existentially fulfilling.

We can, if we like, move farther back in time, seeking additional evidence in the prehistoric evolution of the human species. Clearly, the ability to create tools and weapons helped the earliest manlike creatures to survive. A better spear meant a better diet. The ability to create language also contributed to survival. A clear warning or intelligible instructions could be the difference between life and death. Consequently these creative abilities developed, and man's instinctive urge to use them developed as well. According to the Darwinian theory of natural selection, each tiny change along the way occurred accidentally. Those members of the species who were born with a slightly increased ability and inclination to create had a better chance than their fellows to adapt to their environment and to survive. As the countless generations succeeded each other, brains, hands, eyes—indeed, the entire organism of the average human, became more able and anxious to create—to mold, to carve, to build, and also to devise various means of communication. It is not an exaggeration to say that men are driven to technological creativity because of instincts hardly less basic than hunger and sex.

After organic evolution of the human species was substantially complete, the cultural evolution that had accompanied it continued to do its work. Craftsmanship and invention contribute to survival of the group, so in most societies these activities are prized and encouraged. Thus both genetically and culturally the physical and intellectual traits that contribute to creativity—such as manual dexterity, foresight, and persistence—were developed and reinforced.

There are different types of technological creativity, to be sure. Fabricating a wheel is not the same as inventing it. Construction is not the same as design. Yet all of these human activities are aspects of the creative function, and all have flourished because of the workings of natural selection. Artistic creation is a most complex and marvelous phenomenon. Yet there is evidence that even the artistic impulse has its origins in the inquisitive and analytical traits developed in man by the evolutionary process. Animal survival came first; art and religion came later, byproducts, as it were, of man's technological creativity.

There has lately been some protest against the concept of the species having evolved as *homo faber*, man the maker. It is argued that man developed spiritual and aesthetic capacities independent of, and even prior to, his emergence as a toolmaker. Lewis Mumford, for example, has claimed that early man, instead of being shaped by the struggle for survival, had, inherently, "superorganic demands and aspirations," and that his large brain endowed him with "a tremendous overcharge of psychal energy," which he channeled into rituals, myths, and the like. Man's dreams came first, Mumford tells us, and his inventions followed in due course.

Mumford's theories do not seem convincing to me, and I do not believe that they have gained wide acceptance. The matter does not merit arguing here, since I have no desire to contest the view that man is a dreamer and an artist as well as a technologist. Even the question of which came first, scheming or dreaming, is not of great consequence. The point to be made is that the engineering impulse emerges naturally from our earliest cultures, and even from our genetic constitution.

There are other places in which we can search for our existential heritage. Let us observe a group of children at play. The engineering impulse is clearly present. Let us observe a group of primates. Researchers have spent considerable time demonstrating what has always seemed obvious, "that monkeys have a fundamental curiosity drive, or drive to manipulate."[4]

On this particular search, however, let us not be beguiled by an accumulation of scientific evidence. I can imagine the antitechnologists throwing such evidence back in our faces as another example of single vision, or lack of transcendent energy, or some other such poppycock. In seeking the existential heritage of engineering I propose that we look to the myths of Homer and the Old Testament.

Like the assistants of Hephaistos, engineers have "intelligence in their hearts, . . . and from the immortal gods they have learned to do things." And, like Bezaleel, builder of the tabernacle, the engineer is "filled . . . with the spirit of God, in wisdom, and in understanding, and in knowledge, and in all manner of workmanship."

9. THE EXISTENTIAL ENGINEER

Sisyphus was condemned by the gods to forever roll a huge stone up a mountain, only to see it fall back to the bottom each time he reached the summit. Albert Camus has depicted this mythical figure as the archetypical existential hero. Sisyphus has no illusions to sustain him, no hope that some day his labors will end. But he has pride and courage and the satisfaction that comes instinctively to a person undertaking a task. "The struggle itself toward the heights is enough to fill a man's heart," concludes Camus. "One must imagine Sisyphus happy."

Theodore Roszak has expressed dismay that Camus, "the most great-hearted of our humanist heroes," cannot find in life "a project any less grotesquely absurd than the labor of Sisyphus taking his stone once again up the hill." But Roszak is taking the symbol of the stone too literally. Of course we aim for more than rolling a stone up a hill. But we are beginning to realize that for mankind there will never be a time to rest at the top of the mountain. There will be no new arcadian age. There will always be new burdens, new problems, new failures, new beginnings. And the glory of man is to respond to his harsh fate with zest and ever-renewed effort.

This is why Sisyphus can serve as a symbol of the modern engineer. Today's engineer has lost faith in the utopia that engineers of an earlier age thought they were bringing to mankind. Yet his work, springing as it does from the most basic impulse of humanity, can fill him with existential joy.

That there will be no utopia has become clear beyond questioning. Human beings are too varied, too fickle, and too willful. Technologically oriented optimists like Buckminster Fuller may excite us with visions of glass-domed paradises humming with computers. Humanists like René Dubos may enchant us with tales (featured on *The New York Times* Op-Ed page) of isolated human societies living rich and happy lives under conditions of primitive simplicity. These ideals are interesting, inspiring, and comforting. But we know that they are ideals—perhaps only mirages—that cannot become reality for us. In fact, they are evidence of the differences that surface whenever people start to consider what constitutes the good life.

We have, in our new wisdom and humility, stopped talking about "progress." Except for a few elemental humanitarian concepts, such as not wanting anyone to starve or freeze, we simply cannot agree on which way we want to go. We are talking a lot about "trade-offs," since we have learned that the pursuit of many different worthy objectives results inevitably in conflicts, and that these conflicts can only be resolved by compromise—or by force. The engineer does not underestimate the importance of his contributions to society; but he has abandoned all messianic illusions. He acknowledges that he has made mistakes; but he rejects totally the image of himself as villain, false prophet, or sorcerer's apprentice. He is a human being doing what human beings are created to do: fulfilling his human destiny both biologically and spiritually, and finding his reward in existential pleasure.

This pleasure is not solely the instinctual satisfaction of a lion on the hunt or a beaver building his dam. Pure instinctual gratification is involved, but only as a part of a complex whole. *Homo faber* does not merely putter around, nor is he interested only in survival and comfort. He shares the values and ideals of the human race—mercy, justice, reverence, beauty, and the like. But he feels that these abstract concepts become meaningful only in a world where people lead authentic lives—struggling, questing, and creating.

We have seen the existential pleasures of engineering illuminated by the poetry and myth of Homer and the Old Testament. The question occurs naturally: Where are the poets of the modern era who chant such verses and tell such tales? Where is the poetic evidence that these pleasures still exist? A total absence of artistic verification would not invalidate the claims of the engineer. But such a state of affairs would inevitably make these claims suspect. It is hard to believe that there can exist any valid human experience which does not find its celebration in art.

On this score the engineer need have no apprehensions. The artistic evidence does exist. Perhaps not in voguish movements that occupy the spotlight in fashionable centers of culture. But contemporary creative writers have not been blind to the existential pleasures of engineering. A consideration of this artistic evidence can hardly fail to be a crucial element in our search for a philosophy of engineering to supplant the discredited utopian beliefs of our youth.

For those of us who are engineers, the poetic vision of our profession can serve to reinterpret and refine our own rough-hewn feelings. For young engineers, or would-be engineers, it can give some intimations of the glorious inheritance that is theirs. For those who hold to the view that engineering is soul-deadening and antiexistential (and for those who have never given the matter much thought one way or the other) perhaps the words of poets, novelists and philosophers will open some new and unexpected prospects.

Humanists may be pleased to see us relying upon the creative artist. They may even consider it an embarrassment to us that we find it necessary to look to the poet to give expression to our innermost feelings. But of course we rely upon the artist! He is our cousin, our fellow creator. The artist, in turn, relies upon us. We help to make the world which it is his destiny to interpret. And we give him the materials without which he would be almost totally mute. For the poet we have created pen and ink, the revolutionary printing press, the typewriter, and more lately electronic communications media. For the musician we have created his marvelous instruments and concert halls. ("We have yet sufficiently to realize," according to Lewis Mumford, "that the symphony orchestra is a triumph of engineering.") The history of the fine arts is inextricably intertwined with changing technology, from the introduction of oil paints and lost-wax bronze casting to the cinema and electric arc welding. Right now at the interface between technology and the fine arts there is a frenzy of activity in which it is often difficult to tell the engineers from the artists. In museums and galleries viewers are confronted with restless kinetic sculptures, welded steel forms, flashing neon tubes, inflated plastic shapes, glowing laser beams, pulsing electronic sounds, and a variety of computer-operated multimedia displays. To work with artists—to *become* artists—has been the existential pleasure of several thousand engineers.

Most engineers, of course, are not involved in creating works of art (except as their functional creations may fortuitously emerge as art). But the relationship of engineering to the arts has been one of kinship rather than, as is sometimes suggested, of unrelieved hostility. This is all by way of assuring the humanist that it is no embarrassment for the engineer to look to the creative writer for elucidation of the satisfactions inherent in engineering. Lovers, after all, have never resented Shakespeare for putting into verse those sentiments which can be felt by the ordinary man, but expressed best by the artist.

The first and most obvious existential gratification felt by the engineer stems from his desire to change the world he sees before him. This impulse is not contingent upon the need of mankind for any such changes. Doubtless the impulse was born from the need, but it has taken on a life of its own. Man the creator is by his very nature not satisfied to accept the world as it is. He is driven to change it, to make of it something different. Paul Valéry, in his poetic drama, *Eupalinos*, has expressed this impulse with a romantic flourish:

> The Constructor . . . finds before him as his chaos and as primitive matter, precisely that world-order which the Demiurge wrung from the disorder of the beginning. Nature is formed and the elements are separated; but something enjoins him to consider this work as unfinished, and as requiring to be rehandled and set in motion again for the more special satisfaction of man. He takes as the starting point of his act the very point where the god left off . . . the masses of marble should not remain lifeless within the earth constituting a solid night, nor the cedars and cypresses rest content to come to their end by flame or by rot, when they can be changed into fragrant beams and dazzling furniture.

This desire to change the world is brought to a fever pitch by the inertness of the world as it appears to us, by the very *resistance* of inanimate things, to use the concept expressed by Gaston Bachelard in *La Terre et les Rêveries de la Volonté*:

> The resistant world takes us out of static being. . . . And the mysteries of energy begin. . . .The hammer or the trowel in hand, we are no longer alone, we have an adversary, we have something to do. . . . All these *resistant* objects . . . give us a pretext for mastery and for our energy.

The existential impulse to change the world stirs deep within the engineer. But it is a vague impulse that requires particular projects for its expression. Here the engineer cannot help but be enthralled by the countless possibilities for action that the world presents to him. In *A Family of Engineers,* Robert Louis Stevenson has told of the allure that the profession of engineering had for his grandfather:

> . . . the perpetual need for fresh ingredients stimulated his ingenuity. . . . The seas into which his labours carried the new engineer were still scarce charted, the coast still dark. . . . The joy of my grandfather for his career was as the love of woman.

The engineer today, for all his knowledge and accomplishment, can still look out on seas scarce charted and on coasts still dark. Each new achievement discloses new problems and new possibilities. The allure of these endless vistas bewitches the engineer of every era.

For many engineers, the poetic image of seas and coasts can be taken literally. Water and earth are the substances that engaged the energies of the first engineer—the civil engineer. Civil engineering is the main trunk from which all branches of the profession have sprung. Even in this age of electronics and cybernetics, approximately 16 percent of American engineers are civil engineers. If we add mining, basic metals, and land and sea transportation, fully a quarter of our engineers are engaged in the ancient task of grappling with water and earth.[1] Civil engineering has traditionally included the design and construction of buildings, dams, bridges, railroads, canals, highways, tunnels—in short, all engineered structures—and also the disciplines of hydraulics and sanitation: water supply, flood control, sewage disposal, and so forth. The word "civil" was first used around 1750 by the British engineer, John Smeaton, who wished to distinguish his works (most notably the Eddystone Lighthouse) from those with military purposes. The civil engineer, with his hands literally in the soil, is existentially wedded to the earth, more so than any other man except perhaps the farmer. The civil engineer hero of James A. Michener's novel, *Caravans*, cries out, "I want to stir the earth, fundamentally . . . in the bowels." The hydraulic engineer hero of Dutch novelist A. Den Doolaard's book, *Roll Back the Sea*, stares across the flood water rushing through broken dikes and feels "a strange and bitter joy. This was living water again, which had to be fought against."

Living water. Nature, which appears at one moment to be inert and resistant, something which the engineer is impelled to modify and embellish, in the next instant springs alive as a flood, a landslide, a fire, or an earthquake, becomes a force with which the engineer must reckon. Beyond emergencies and disasters, through the environmental crisis of recent years, nature has demonstrated that she is indeed a living organism not to be tampered with unthinkingly. Nature's apparent passivity, like the repose of a languid mistress, obscures a mysterious and provocative energy. The engineer's new knowledge of nature's complexities is at once humbling and alluring.

Another dichotomy with which nature confronts the engineer relates to size. When man considers his place in the natural world, his first reaction is one of awe. He is so small, while the mountains, valleys and oceans are so immense. He is intimidated. But at the same instant he is inspired. The grand scale of the world invites him to conceive colossal works. In pursuing such works, he has often shown a lack of aesthetic sensibility. He has been vain, building useless pyramids, and foolish, building dams that do more harm than good. But the existential impulse to create enormous structures remains, even after he has been chastened. Skyscrapers, bridges, dams, aqueducts, tunnels—these mammoth undertakings appeal to a human passion that appears to be inextinguishable. Jean-Jacques Rousseau, the quintessential lover of nature undefiled, found himself under the spell of this passion when he came upon an enormous Roman aqueduct:

> I walked along the three stages of this superb construction, with a respect that made me almost shrink from treading on it. The echo of my footsteps under the immense arches made me think I could hear the strong voices of the men who had built it. I felt lost like an insect in the immensity of the work. I felt, along with the sense of my own littleness, something nevertheless which seemed to elevate my soul; I said to myself with a sigh: "Oh! that I had been born a Roman!" . . . I

> remained several hours in this rapture of contemplation. I came away from it in a kind of dream. . . .

The rapture of Rousseau for "the immensity of the work" survives in the midst of our most bitter disappointments with technology. A 1964 photo exhibition at The Museum of Modern Art in New York, entitled "Twentieth-Century Engineering," brought home this truth to a multitude of viewers. The introduction to the exhibition catalogue directed attention to the fact that the impact of enormous engineering works is sometimes enhanced by the "elegance, lightness, and the apparent ease with which difficulties are overcome," and sometimes by the opposite, the monumental extravagance that appears when "the engineer may glory in the sheer effort his work involves." Ada Louise Huxtable of the *New York Times* reacted to the show with an enthusiasm that even the proudest of civil engineers would hesitate to express:

> It is clear that in the whole range of our complex culture, with its self-conscious aesthetic kicks and esoteric pursuit of meanings, nothing comes off with quite the validity, reality, and necessity of the structural arts.
>
> Other art forms seem pretty piddling next to dams that challenge mountains, roads that leap chasms, and domes that span miles. The kicks here are for real. These structures stand in positive, creative contrast to the willful negativism and transient novelty that have made so much painting and literature, for example, a kind of diminishing, naughty game. The evidence is incontrovertible: building is the great art of our time.

"The kicks here are for real." And if they are for real to the observer of photographs, imagine what they are like to the men who participate in creating the works themselves. *Roll Back the Sea*, the Dutch novel already mentioned, has a scene describing the building of the Zuyder Zee wall which gives some slight taste of the excitement surrounding a massive engineering work:

> The great floating cranes, dropping tons of stiff clay into the splashing water with each swing of the arm. Dozens of

> tugboats with their white bow waves. Creaking bucket dredges; unwieldy barges; blowers spouting the white mass of sand through long pipelines out behind the dark clay dam; and the hundreds of polder workmen in their high, muddy boots. An atmosphere of drawing boards and tide tables, of megaphones and jingling telephones; of pitching lights in the darkness, of sweat and steam and rust and water, of the slick clay and the wind. A dike in the making, the greatest dike that the world had ever seen built straight through sea water.

The mighty works of the civil engineer sometimes appear to be conquests over a nature that would repel mankind if it could. Thus Waldo Frank perceived the Panama Canal slashing through the tropical jungle:

> Its gray sobriety is apart from the luxuriance of nature. Its wilfulness is victor over a voluptuary world that will lift no vessels, that would bar all vessels.

At other times the civil engineer's structures appear to grow out of the earth with a natural grace that implies the fulfillment of an organic plan. Pierre Boulle, in *The Bridge Over the River Kwai*, writes: "An observer, blind to elementary detail but keen on general principles, might have regarded the development of the bridge as an uninterrupted process of natural growth." The bridge rose day by day, "majestically registering in all three dimensions the palpable shape of creation at the foot of these wild Siamese mountains. . . ." Fifty years after the construction of the Eiffel Tower a Parisian recalled: "It appeared as if the tower was pushing itself upward by a supernatural force, like a tree growing beyond bounds yet steadily growing. . . . Astonished Paris saw rising on its own grounds this new shape of a new adventure."

From the organic implications of the civil engineer's structures it is but a short step to the spiritual. Mighty works of concrete, steel, or stone, seeming alive but superhuman in scope, inevitably invoke thoughts of the divine. The ultimate material expressions of religious faith are, of course, the medieval cathedrals. They are usually defined as the material creations of religious men. But they can also be considered as magnificent works of engineering which, through their physical majesty and proportion, impel the viewer to think lofty thoughts. In

Mont Saint-Michel and Chartres Henry Adams has conveyed a sense of the way in which these physical structures both reflect and evoke a spiritual concept:

> Every inch of material, up and down, from crypt to vault, from man to God, from the universe to the atom, had its task, giving support where support was needed, or weight where concentration was felt, but always with the condition of showing conspicuously to the eye the great lines which led to unity and the curves which controlled divergence; so that, from the cross on the fleche and the keystone of the vault, down through the ribbed nervures, the columns, the windows, to the foundation of the flying buttresses far beyond the walls, one idea controlled every line.

William Golding, in his novel *The Spire*, has explored the theme of the interrelationship between construction and religion. Set in medieval England, the novel relates the story of the building of a cathedral tower, a tower which threatens to cause the collapse of the structure on which it rests. Priest and master builder confront each other, and the construction is accompanied by their dialogue, the dialogue between faith and technology. At one point the priest addresses the master builder in these words:

> My son. The building is a diagram of prayer; and our spire will be a diagram of the highest prayer of all. God revealed it to me in a vision, his unprofitable servant. He chose me. He chooses you, to fill the diagram with glass and iron and stone, since the children of men require a thing to look at. D'you think you can escape? You're not in my net—. . . It's His. We can neither of us avoid this work. And there's another thing. I've begun to see how we can't understand it either, since each new foot reveals a new effect, a new purpose.

Not only cathedrals, but every great engineering work is an expression of motivation and of purpose which cannot be divorced from religious implications. This truth provides the engineer with what many would assert to be the ultimate existential experience.

I do not want to get carried away on this point. The age of cathedral building is long past. And, as I have already said, less than one-quarter of today's engineers are engaged in construction activities of any sort. But every manmade structure, no matter how mundane, has a little bit of cathedral in it, since man cannot help but transcend himself as soon as he begins to design and construct. As the priest of *The Spire* expresses it: "each new foot reveals a new effect, a new purpose."

In spite of the many ugly and tasteless structures that mar our cities and landscapes, public enthusiasm for building has survived relatively unscathed through the recent years of disenchantment with technology. The engineer, in company with architects, artists, and city planners, has kept alive the public faith in the potentiality for beauty, majesty, and spirituality in construction.

At a time when we are embarrassed to recall the grandiose pronouncements of so many of our predecessors, the First Proclamation of the Weimar Bauhaus, dating from 1919, retains its dignity and ability to inspire. It was the concept of architect Walter Gropius that great art in building grew out of craftsmanship, was in fact nothing other than craftsmanship inspired. His concept of craftsmanship included necessarily the essentials of civil engineering. "We must all turn to the crafts," he told his followers:

> Art is not a "profession." There is no essential difference between the artist and the craftsman. The artist is an exalted craftsman. In rare moments of inspiration, moments beyond the control of his will, the grace of heaven may cause his work to blossom into art. *But proficiency in his craft is essential to every artist.* Therein lies a source of creative imagination.
>
> Let us create a *new guild of craftsmen,* without the class distinctions which raise an arrogant barrier between craftsman and artist. Together let us conceive and create the new building of the future, which . . . will rise one day toward heaven from the hands of a million workers like the crystal symbol of a new faith.

Enough, then, of the civil engineer and his wrestling with the elements, his love affair with nature, his yearning for immensity, his raising toward heaven the crystal symbol of a new faith. His existential bond to the earth, and expression of his own elemental being, need no further amplification, no additional testimonials.

10. "LOOK LONG ON AN ENGINE. IT IS SWEET TO THE EYES."

The time has come to confront a more difficult subject: the relationship of the engineer to the central protagonist in the history of technology, the principal villain in the current environmental crisis, the symbol and enigma of our age—the machine. It has been said that we have already entered a postmachine age; but this is one of those statements better suited to catching the eye than revealing the truth. It is a fact that some engineers are turning their attention away from mechanical problems to the theoretical study of systems, yet statistically they still form a very small percentage of the profession. It is also a fact that many engineers today are working with computers and electronic devices, which surely make less noise than lathes and power presses, and perhaps are less "mechanical," but qualify as machines nevertheless. Although accurate and meaningful statistics are hard to come by, it can be said with some assurance that at least half of America's million or so engineers are mechanical, electrical, or aerospace specialists who are intimately engaged with machines.[1]

Mankind has lived with mechanical devices since the invention of the wheel and other labor-saving contrivances of prehistory. But *The Machine* has only been with us during the two centuries that have passed since the beginning of the industrial revolution. The rapid development and dispersion of the machine was inextricably combined with the devlopment of new sources of power, first steam, then internal combustion, electricity, and finally the atom. The machine is what served as the engineer's magic wand. It is what made it possible for him to transform the world, mesmerize the populace, and make of himself a folk hero and prospective savior. It (along with some chemical wizardry) is what brought about the environmental crisis, engendered a sudden, new hostility toward the engineer, and sparked the emergence of antitechnological sentiment throughout the land.

How, at this point, is the engineer to relate to this phenomenon—his creation, his child, his servant, his master, his glory, and his nemesis?

If he seeks guidance from the creative writer, he finds himself confronted with such masses of machine-inspired poetry and prose that he is hard-pressed to know where to look first. Happily there are some useful studies that can serve as an introduction and guide. Peter Viereck, for example, in an article entitled "The Poet in the Machine Age," has classified writers in four categories of "anti-machine," and an equal number of "pro-machine." The "antis" he lists as (1) the "aesthetic wincers," including the English Lake poets and the French Symbolists, to whom the machine represents dirt, ugliness, and vulgarity; (2) the "pious scorners," including such concerned Catholics as G. K. Chesterton and Hilaire Belloc, who feared the moral consequences of the passing of traditional rural life; (3) the "back-to-instinct prophets," such as D. H. Lawrence, and more recently Henry Miller, who defend the impulses of the human heart against the "artificiality" of the machine; and finally (4) the "trapped individualists," who fear that in a mechanized world men inevitably must lose their uniqueness, their free will, what we might now call their existential essence. A charter member of this latter group was Matthew Arnold, who bewailed the fact that England

> Stupidly travels her round
> Of mechanic business and lets
> Slow die out of her life
> Glory, and genius, and joy.

The contemporary antitechnologists, as we have seen, find inspiration in each of these four sources.

The four categories of pro-machine poets Viereck lists as (1) the "middle-class materialists," and (2) the "socialist materialists," mostly writers of atrocious poetry who, in the early days of the industrial revolution, celebrated the machine as the redeemer of man from arduous toil, the first group for the benefit of the stability of the realm, the second for the benefit of the working classes. Third, Viereck mentions the "gadget cultists," who delight in the machine as an aesthetically attractive object. Kipling must be included in this group, although he certainly shared the empire-building sentiments of the middle-class materialists. Fourth come the "lion-tamers," such as Ralph Waldo Emerson, those who grudgingly accept the presence of the machine, but dream of it being kept under control for the ethical and aesthetic benefit of humanity.

Other studies suggest different avenues of thought. Herbert L. Sussman, in *Victorians and the Machine*, notes that English writers, in their early encounters with the machine—and to a certain extent, ever since—have avoided confronting it as an object in itself, but regarded it as a transcendental force, symbolizing spiritual principles and moral qualities, sometimes a godlike benevolence, often a devilish depravity. Frequently machines have been depicted metaphorically as rearing steeds, roaring bulls, or hissing dragons. In *The Poet and the Machine,* Paul Ginestier, a French writer drawing mostly on European sources, remarks that the machine has relieved some of mankind's deeply seated feelings of impotence, and that the poet rightfully celebrates this occurrence. However, according to Ginestier, the poet of the machine, like a psychiatrist, plays a double role, reminding us of the bounds of reality at the same time that he exalts our achievements. As seen by Thomas Reed West, author of *Flesh of Steel, Literature and the Machine in American Culture*, machines impose disciplines on human spontaneity and freedom. It is West's interesting insight that these disciplines "suggest an ascetic strength rather than a cultural decline," but he acknowledges that most contemporary authors reject such a view.

Leo Marx, in *The Machine in the Garden*, considers the response of American writers to the invasion of the machine into the landscape of the New World. This response, Marx reminds us, must be viewed against the concept of America as a virgin land, a pastoral Eden in which the roaring locomotive—the machine most frequently depicted—evoked apprehensions, and occasionally aspirations, of a particular intensity.

Much creative writing about the machine is of limited interest to the engineer who seeks artistic interpretation of his own experience. Poems and stories that are hostile to the machine are either antiquated or foolish, or else express a message that the engineer has already heard a thousand times. We do not need to be chastised again and again about the ugliness of "satanic mills," and about the evils of soot and automobile accidents. Barbara Howes's poem, "Headlong," which describes a car running over a woodchuck in Vermont, ends with these words:

> This headlong meeting stopped us: I ran
> A gauntlet of chill air the long way home.

Engineers by now have run their "gauntlet of chill air," and are very much aware of the potential malignity of machines. This awareness—coming from statistics and public opinion, as well as from art—is the basis of our new humility. When we think of machines today we are implicitly thinking of machines that do a maximum of "good" and a minimum of "evil." Poetry and prose hostile to machines does exist; much of it has merit and surely deserves an audience. But in our present search it does not offer us either enticement or reward. We have acknowledged our mistakes. At this moment we are seeking to define those aspects of our profession which are most worthy and most fulfilling.

When we turn to the writers who are admirers of the machine, the problem is to find among them a quality of thought and expression that does not lapse into triteness and triviality.

The engineer's first instinctive feeling about the machine is likely to be a flush of pride. For all the mistakes that have been made in its use, the machine still stands as one of mankind's most notable achievements. Man is weak, and yet the machine is incredibly strong and productive. The primordial joy of the successful hunt or the abundant harvest has its modern counterpart in the exhilaration of the man who has invented or produced a successful machine.

This pride, unfortunately, does not translate readily into great literature. It is small comfort to the engineer to read the impassioned verse of a Mr. Ebenezer Elliott:

> Engine of Watt! unrivall'd is thy sway.
> Compared with thine, what is the tyrant's power?

His might destroys, while thine creates and saves.
Thy triumphs live and grow, like fruit and flower . . .

More than two centuries of such doggerel have helped to confirm the machine's bad reputation among people of sensitivity.

There is even less comfort—indeed there is alarm—in reading the proud proclamations of the Futurists, a group of Italian intellectuals and artists who in the early years of this century boasted that the coming of machines would replace the stagnant traditions of their nation with a glorious new order. "A great pride swelled in our chests," wrote Filippo Tommaso Marinetti in his introduction to the first Futurist Manifesto. But this great pride seemed to stem mostly from his ability to leap into a motor car, a "snorting beast," and go careening through the streets at breakneck speed. A love of power, speed, and danger were the distinctive features of Futurism, followed by an impetuous glorification of war and fascism. Pride in the machine is obviously an emotion that has its dark side.

However, an awareness of the trite and the reckless need not stifle the legitimate and wholesome self-esteem that the engineer feels in his mechanical creations. For reassurance, he can turn to William Wordsworth, whose credentials as an artist of sensibility are impeccable, and yet who could not help but express his delight in the marvel of machinery:

. . . . yet do I exult,
Casting reserve away, exult to see
An intellectual mastery exercised
O'er the blind elements; a purpose given,
A perseverance fed; almost a soul
Imparted—to brute matter. I rejoice,
Measuring the force of those gigantic powers
That, by the thinking mind, have been compelled
To serve the will of feeble-bodied Man.

The verse may not be Wordsworth's most inspired, and the pride is tempered later in the poem by the hope that "this dominion over nature" will be exercised with due respect for the moral law. But, grudging or not, the great poet's admiration is there to bear witness to the validity of the engineer's exultation.

After the engineer's initial burst of pride has run its course, quite a different sentiment reveals itself—his love of the machine for its intrinsic beauty. The machine, as experienced by the engineer, is a marvel of dynamic elegance—not a copy of a horse or a dragon as some poets have tried to depict it, but in its own essence. Rudyard Kipling and a few of his contemporaries tried to express this truth in verse, although with limited success. Perhaps the most famous poem in this genre is Kipling's "M'Andrews Hymn," a few lines of which serve to give the flavor of the whole:

> The crank-throws give the double-bass, the
> feed-pump sobs an' heaves,
> An' now the main eccentrics start their quarrel
> on the sheaves:
> Her time, her own appointed time, the rocking
> link-head bides,
> Till—hear that note?—the rod's return whings
> glimmerin' through the guides.
> They're all awa! True beat, full power, the
> clangin' chorus goes
> Clear to the tunnel where they sit, my purrin'
> dynamoes.

Walt Whitman embraced the dynamic beauty of machinery as he embraced just about every aspect of life, and expressed his admiration for the archetypical machine of his day, the locomotive:

> Fierce-throated beauty!
> Roll through my chant with all thy lawless
> music, thy swinging lamps at night,
> Thy madly-whistled laughter, echoing,
> rumbling like an earthquake, rousing all,
> Law of thyself complete, thine own track
> firmly holding . . .

In our own time Stephen Spender has worked with the same theme. These lines are from "The Express":

Ah, like a comet through flame, she moves
entranced,
Wrapt in her music no bird song, no, nor
bough
Breaking with honey buds, shall ever equal.

Spender, perhaps more than any other contemporary poet, has had success with integrating the beauty of the machine into poetry of merit. This portion from "The Landscape Near an Aerodrome" is a well-known example:

More beautiful and soft than any moth
With burring furred antennae feeling its
huge path
Through dusk, the air liner with shut-off
engines
Glides over suburbs and the sleeves set trailing
tall
To point the wind. Gently, broadly, she falls,
Scarcely disturbing charted currents of air.

MacKnight Black, a lesser known poet, is the author of an entire book of verse entitled *Machinery*, some of which cannot help but strike a responsive chord in the engineer:

The arc of a balance-wheel
Flows like a curved rush of swallows, come
over a hill. . . .
Things lost come again in sudden new beauty.
Look long on an engine. It is sweet to the
eyes.

Aldous Huxley has said that men of letters are interested in the social and psychological consequences of technology, but not in "the embodied logic of machinery" itself. If this is true, it would help to explain why so little important literature is devoted to the beauty and grandeur of the machine. Undeniably what Herbert L. Sussman has observed about Victorian writers seems to have prevailed right to the present day: "With very few exceptions . . . the machine appears in the minor works of major poets or the major works of minor poets." But perhaps it is merely a case of the great poet of machinery not yet having arrived on the scene. In the Kipling poem quoted above, M'Andrews calls out, "Lord, send a man like Robbie Burns to sing the Song o' Steam!" A machine-loving Robbie Burns may yet appear among us at any moment.

If there is a sparseness of talented poets and novelists extolling the beauty of machinery, the ranks are more than filled by spokesmen from the realm of the fine arts. Francis Picabia, famous along with Marcel Duchamp for the introduction of machines as central subjects for painting, announced in 1915 that "the genius of the modern world is in machinery and that through machinery art ought to find a most vivid expression." To artists of this school the machine is not an alien intruder into the organic world, but a natural part of it, a part also of man's inner visionary landscape. A review published in Paris by followers of Picabia declared that to an artist of this sort

> . . . the world of ideas and forms appears like a sympathetic cosmos filled with correspondences, relationships, and resemblances. He perceives what may be the common link between a flower and a combustion engine, between a line and an idea, a color and a memory, a love and a chemical phenomenon, a biblical personage and a doctrine of art, a piano and a comb, the sea and a streetcar. . . . His only objective is to trust, to project into material form the realities of his inner self. So every work of art becomes the representation of a private world. . . .

Other artists have found in the machine a pure beauty that seems completely isolated from subjective human experience. Fernand Léger tells of visiting an airplane exhibition with fellow artists Duchamp and Brancusi. Duchamp, according to the story, turned to Brancusi and said, "Painting has come to an end. Who can do anything better than this propellor?" "I myself," relates Léger, "felt a preference for the motors. . . . But I still remember the bearing of those great propellers. Good God, what a miracle!"

For the engineer there is ample evidence both from men of letters and men of the fine arts that the machine, rather than cutting him off from the wellsprings of beauty, opens before him new vistas of truth, splendor, and elegance. The beauty of the machine is pure, like mathematics. It is also, paradoxically, imbued with the vitality of humanity, since it is exclusively man-made. Freud regarded many manufactured articles as sexual symbols, and several artists have seen the machine in the same way.

Pure and complex, mechanistic and organic, tranquil and violent, objective and subjective—in all its enigmatic complexity, the machine has added to the richness of man's aesthetic experience. "The machine has not separated us from nature," observed the Russian painter El Lissitzky. "Through the machine we have discovered a new nature which previously was not envisioned."

This is not to deny that machines have contributed their share of ugliness to the world. Yet such ugliness often has a strangely transient quality about it. Machines tend to age gracefully. Even in an era of disenchantment with technology, intimacy serves to enhance the attractiveness of the machine rather than to diminish it. A patina of familiarity softens its surfaces and enhances its appeal. The prime example is what has happened to the locomotive. Once regarded as an ugly monster, its beauty and charm are now widely acclaimed.

As machines become more physically attractive with the passage of time, they also have a way of becoming more cordial. Siegfried Sassoon, sitting alone at night writing poetry, hears a train in the valley and concludes his verse, " 'That train's quite like an old familiar friend,' one feels." Antoine de Saint Exupéry has said the same thing about the locomotive, which was once greeted with such trepidation: "What is it today for the villager except a humble friend who calls every evening at six?" Saint Exupéry sees the same softening and accommodation occurring to each machine in its turn.

The machine that was once austere, or even ungainly, slowly, mysteriously becomes an attractive companion. There is a simple, quiet joy that an engineer can feel in the presence of a machine that is equivalent in a way to the reverie that can come upon a naturalist in a forest setting. In the early pages of his novel, *The Sand Pebbles*, Richard McKenna attempts to evoke this mood:

> Holman sat on the workbench, careless of oil on his dress whites. The spell of the engine was on him.
>
> It was a fine, handsome old engine, much older than Jake Holman himself. He looked at it, massive, dully gleaming brass and steel in columns and rods and links arching above drive rods from twinned eccentrics, great crossheads hung midway, and above them valve spindles and piston rods disappearing into the cylinder block. He knew them all, each part and its place in the whole, and his eye followed the pattern, three times repeated from forward to aft, each one-third of the circle out of phase, and it was all poised and balanced there like three chunks of frozen music. Under his controlling hands, when they steamed, it was going to become living, speaking music. Under his tending hands, with oil can and grease swab. Under his healing hands, with hammer and wrench and scraper.
>
> "Hello, engine. I'm Jake Holman," he said under his breath.
>
> Jake Holman loved machinery in the way some other men loved God, women and their country. He loved main engines most of all, because they were the deep heart and power center of any ship and all the rest was trimming, much of it useless. He sat and looked at the engine without thinking. . . .

Here I had better interject very quickly the reminder that professional engineers are not to be confused with mechanics or technicians. Yet, in considering the emotional relationship of man to machine, the engineer shares a common bond with every fellow being who works with machines, understands them, and is entranced by them.

Does it sound odd to say that machinery can be comforting, can actually become cosy? Hear Anne Morrow Lindbergh in *Listen the Wind!*, describe her feelings about an airplane:

> This little cockpit of mine became extraordinarily pleasing to me, as much so as a furnished study at home. Every corner, every crack, had significance. Every object meant something. Not only the tools I was working with, the transmitter and receiver, the key and the antenna reel; but even the small irrelevant objects on the side of the fuselage, the little

> black hooded light, its face now turned away from me, the shining arm and knob of the second throttle, the bright switches and handles, the colored wires and copper pipes: all gave me, in a strange sense, as much pleasure as my familiar books and pictures might at home. The pleasure was perhaps not esthetic but came from a sense of familiarity, security, and possession. I invested them with an emotional significance of their own, since they had been through so much with me. They made up this comfortable, familiar, tidy, compact world that was mine. . . . Outside the night rushed by. How nice to be in your own little room, to pull your belongings around you, to draw in like a snail in his shell, to work!

Every engineer has experienced the comfort that comes with total absorption in a mechanical environment. The world becomes reduced and manageable, controlled and unchaotic. For a period of time, personal concerns, particularly petty concerns, are forgotten, as the mind becomes enchanted with the patterns of an orderly and circumscribed scene. This state of mind is scorned by many humanists, but in a way it is similar to the comfortable seclusion one feels when listening to a carefully constructed musical composition of the classical period. If such a state of mind comes to dominate one's life, then of course it can be said to be dehumanizing. But its absence from life deprives the individual of that "getting-out-of-himself" which is an important part of the human adventure. Philosophers and religious thinkers are constantly talking about "losing oneself" in the All, and then in the next breath of "finding oneself" in some form of ecstasy. These emotional conditions are difficult to define verbally with any precision. But somewhere among the states of being sought by wise men falls that wondrous moment in which the engineer becomes absorbed with the machine.

Contemporary Russian novelists seem to have a particular feel for this experience, perhaps because their attitude toward technology has not turned as sour as it has in many other lands. Vladimir Dudintsev, in *Not by Bread Alone*, sends his engineer hero out for a walk:

> From his first few paces after leaving the house, Dmitri forgot everything; his soul left his body and flew to a world of machines, while his legs worked automatically. . . .

Andrey Platonov, in his story "Fro," conceives of his engineer "entering into the very essence of the abstruse, inanimate mechanisms. . . ."

> He had the faculty of actually feeling the degree of intensity of an electric current, as if it were a secret passion of his own. . . . he had acquired a true conception of the forces at work in any mechanical setup and an immediate insight into the strength of endurance of the basic metal in any machine.

An imprisoned engineer in Aleksandr Solzhenitsyn's *The First Circle* finds a form of freedom in his preoccupation with electronic equipment:

> . . . hardly had he entered the laboratory that morning than the inner logic of work took hold of him, suppressing all other feelings and thought. This capacity to devote himself wholly to his work, to forget about life, had been the basis of his engineering triumphs on the outside, and in prison it helped him bear his misfortune.

It is crucial that the engineer remember that the "forgetting about life" is only for certain select periods of time, and must not be allowed to become a permanent condition. The joy of engrossment in the mechanical, like all of the existential pleasures of life, has the potentiality of becoming a destructive obsession. In *The Builder of Bridges*, a long-forgotten play that graced the London stage during the first decade of this century, Edward Thursfield, resident engineer, tells his boss, Sir Henry, that he does not wish to go to Rhodesia on an engineering project because he is planning to get married:

> EDWARD: I'm in love, Chief.
>
> SIR HENRY: So was I, once, but I shut myself up for a week, and worked at an air-machine. Grew so excited I forgot the girl. You try.

Happily, as was demonstrated by the denouement of the play, the excitement of working with machines need not preclude other excitements, including what used to be called affairs of the heart.

In his emotional involvement with the machine, the engineer cannot help but feel at times that he has come face to face with a strange but potent form of life. We have spoken already of the "life" that the civil engineer encounters in the forces of nature. But with machines the situation is quite different. A manufactured device is obviously more "artificial" than a river, and yet it can seem more alive. First of all, as poets keep reminding us, machines look like living things. "The windmills, like great sunflowers on dried stalks,/Stare hard at the sun they cannot follow."[2] They sound like living things. A turbine "hums there softly, purring with delight."[3] They move like living things; take Longfellow's ship:

> And see! she stirs!
> She starts—she moves—she seems to feel
> The thrill of life along her keel,
> And, spurning with her foot the ground,
> With one exulting, joyous bound,
> She leaps into the ocean's arms!

They throb with power, as felt by the pilot of Saint-Exupéry's *Night Flight*:

> He passed his fingers along a steel rib and felt the stream of life that flowed in it; the metal did not vibrate, yet it was alive. The engine's five-hundred horse-power bred in its texture a very gentle current, fraying its ice-cold rind into a velvety bloom. Once again the pilot in full flight experienced neither giddiness nor any thrill; only the mystery of metal turned to living flesh.

And as the "mechanical" machines, which remind us of muscles, bones, and circulatory systems, are supplemented by an increasing number of electronic and chemical devices, which imitate the processes of the brain and nervous system, the sense of vitality in the machine increases accordingly.

When we speak of sensing life in the machine, we cannot literally conceive of a robot touched by a magic wand which quickens it into a living creature. We can think only in terms of the basic energy of the universe, which physicists (and philosophers) have told us is also matter—one pulsing essence which is manifest everywhere. We encounter "it" in forests, of course, and in clouds, and within ourselves. But as engineers, we also encounter "it" repeatedly in the machines we design and manufacture and use. This intensifies our existential experience of the world. I have used the word "it" to identify the basic energy-matter. There are other more graceful and meaningful terms that can serve. Robert M. Pirsig, in his intriguing book, *Zen and the Art of Motorcycle Maintenance*, expresses the thought this way:

> The Buddha, the Godhead, resides quite as comfortably in the circuits of a digital computer or the gears of a cycle transmission as he does at the top of a mountain or in the petals of a flower. To think otherwise is to demean the Buddha—which is to demean oneself.

To be totally immersed in the circuits of a computer, or the gears of a transmission, according to Pirsig, is to have access to that very peace of mind which is so often said to be unattainable to practitioners of technology. To follow Pirsig a little further on this, engineering work is seen to be a means of acquiring an inner peace of mind on three levels: physical quietness; mental quietness, "in which one has no wandering thoughts . . ."; and value quietness, "in which one has no wandering desires at all but simply performs the acts of his life without desire . . ." This peace of mind is not only a consequence of working with machines, it is a means toward performing such work successfully, since it "seems to draw out the inner tensions and frustrations that have prevented you from solving problems you couldn't solve before. . . ."

11. "THEN WAS I CARRIED BEYOND PLEASURE."

I do not want to wander too far into the realm of mystic experience, since engineering is an earthy profession firmly based in the secular desires and activities of men. Yet we cannot help but be drawn in this direction. The engineer, after all, spends much of his time intimately involved with those eternal principles which we call "the laws of nature." It is possible, I suppose, to look at the natural laws—the symmetries, proportions, and regularities of the universe—with a cold and analytical eye. But even the dullest of us is moved to wonder by contemplation of the universal order. The romantic poets themselves, for all their apprehensions about science, cannot resist expressing their awe. From meditating on Nature's laws Wordsworth found "a pleasure quiet and profound" and the "comfort" of a "transcendent peace."

From a consideration of the machine, we have arrived inevitably at the doorstep of science. The work and objectives of the engineer are quite different from those of the scientist. But on occasion the engineer shares in the scientist's satisfactions. When Einstein, for example, speaks of the "great eternal riddle" of the universe, and how he and fellow scientists "found inner freedom and security in devoted occupation with it," the engineer can readily empathize. Quasi-mystical moments of peace and wonder are not inconsistent with the engineer's other moods—his enthusiasm, his pride, his craving for action and practical accomplishment. Arthur Koestler has written about finding in the theories of Einstein "a sensation of infinite tranquility and peace" which overwhelmed for a time—but only for a time—his concern and anger about the fearful condition of world affairs. The engineer, in oscillating between action and tranquility, creation and contemplation, realizes to the fullest extent his existential potential.

Beyond the world of science, the engineer encounters the pristine realm of pure mathematics. Although for the engineer mathematics is a tool rather than an end in itself, he has the opportunity to partake of its serene gratifications. Once more let us call on Wordsworth to attest to the pleasures we feel:

> Mighty is the charm
> Of these abstractions to a mind beset
> With images, and haunted by herself,
> And specially delightful unto me
> Was that clear synthesis built up aloft
> So gracefully.

However delightful we may find the "clear synthesis built up aloft," and however well we may understand Edna St. Vincent Millay's line that "Euclid alone has looked on beauty bare," we cannot stay too long in the rarefied atmosphere of mathematical theory. We can be thankful that the pure beauty of mathematics is part of our life experience, but we must remember that our proper place is on the mainland of "doing."

To the engineer, however, "doing" is something more than mere manufacturing. An essential element of the profession of engineering is the concept of *creativity.* Implicit in all previous discussion of the different types of engineering activity—civil, mechanical, electrical, chemical, and the rest—is the assumption that engineers design things. Indeed, most engineers do design things. But the word design covers a tremendous variety of functions, ranging from the simplest calculation of the size of a beam required to support a given load on a given span, up to the invention of the transistor, or the development of a vehicle to land on the moon. Singly and in groups, from the commonplace to the ingenious, most engineers have some opportunity to create. Only about one-third of them are engaged directly in design, development and research.[1] Another third are in what they call "planning, directing, advising, consulting and teaching," so we can assume that most of these have creative opportunities. As for the last third—selling, producing, constructing, maintaining, testing, and so forth—it is hard to tell. But, however it works out statistically, creative design is the central mission of the professional engineer. In a world where so much energy is devoted to buying and selling, this is his delight as well as his contribution.

The existential response to a successful design, creation, discovery, or invention, in science or in engineering, can range from calm satisfaction to absolute rapture. C. P. Snow, in his novel *The Search*, describes the thrill of the verification of a scientific prediction on which the narrator has been working for a long time:

> Then I was carried beyond pleasure. . . . My own triumph and delight and success were there, but they seemed insignificant beside this tranquil ecstasy. It was as though I had looked for a truth outside myself, and finding it had become for a moment part of the truth I sought; as though all the world, the atoms and the stars, were wonderfully clear and close to me, and I to them, so that we were part of a lucidity more tremendous than any mystery.
>
> I had never known that such a moment could exist. . . .
>
> Since then I have never quite regained it. But one effect will stay with me as long as I live; once, when I was young, I used to sneer at the mystics who have described the experience of being at one with God and part of the unity of things. After that afternoon, I did not want to laugh again; for though I should have interpreted the experience differently, I thought I knew what they meant.

This is an account of existential pleasure indeed! For the engineer, however, as opposed to the scientist, the fullest gratification is reserved for that creative solution which achieves a desired practical result. As an example of this ultimate experience let us consider a haunting and provocative novel, *The Woman in the Dunes*, by the noted Japanese author Kobo Abé. A young man, vaguely dissatisfied with the routine of his life in the city, goes on a weekend excursion to an isolated beach. Here he becomes trapped at the bottom of a huge sand pit in which there is a house occupied by a single woman. He is informed that unless he assists her in perpetually digging sand, which is hauled away nightly by the people from a nearby village, the pit will fill with sand, her house will be inundated, and the dunes will advance to overwhelm the next house, and eventually the entire village. The villagers force him to assist in the digging by withholding drinking water from him when he does not comply with their orders. There is no possible escape. As he comes to recognize the hopelessness of his situation, he sinks into a deep melancholy. The symbolic representation of our contemporary situation is clear.

One day the protagonist finds that he can obtain fresh water from the sand with a simple device based on the principle of capillary action. This discovery has a remarkable effect upon him:

> He had to sit down for a moment and control his breathing in order to quiet the wild beating of his heart. . . .
>
> But he could not suppress the natural laughter that welled up in him. . . . It was as if his stomach were being tickled by a paper balloon filled with some special light gas. He felt that the hand he held to his face was floating free in the air. . . .
>
> He was thinking of the vast network of water veins creeping up through the sand. . . .
>
> The fact that he was still just as much at the bottom of the hole as ever had not changed, but he felt quite as if he had climbed to the top of a high tower.

The young man, by the exercise of his ingenuity, has devised a solution that enables him to achieve some control over his own destiny. The technical accomplishment delights him. But it is the new freedom from the domination by alien forces that makes his satisfaction complete. Given an opportunity to escape, he resolves to remain. He is anxious to improve his device, and contentedly contemplates new calculations and experiments. He is anxious to share the knowledge of his discovery with others. He has become attached to his woman companion. His life in the "outside" world had offered him no great satisfaction in the first place. The author's conclusion is enigmatic. The young man is "trapped." Yet we are all trapped. Attempts to escape the human condition are doomed to failure. To cope creatively with our environment, to help our fellow humans survive with dignity, to undertake necessary tasks with courage and imagination, is to live out our destiny, and fulfill our existential yearnings. As Camus says of Sisyphus, so can we say of the young man: "One must imagine him happy."

There remains one more existential gratification which is experienced by the engineer. This—the final and the most important of all—stems from his contribution to the welfare of society. It may seem paradoxical to make such a statement after all that has been said up to this point about the failure of technology to make men happy, and about the foolish hopes held by engineers of earlier days. But despite the fact that most engineers have become acutely aware of the disagreeable problems inherent in technological change, and have relinquished all illusions of redeeming mankind, there still exists a strong sense of ***helping,*** of directing efforts toward easing the lot of one's fellows. Santayana has observed that the religion we return to in our mature years is not the same as the one we become estranged from in our youth. By analogy, the feeling of contribution that the engineer has today is far different from the arrogant confidence of the profession's earlier years. If we are less sure that we are making men happier, we are at least buoyed by the knowledge that we are trying.

Goethe's Faust, jaded with every conceivable worldly experience, finally found—in a land-reclamation project—the contentment that had eluded him all his life. A literary critic has commented that to the contemporary reader "it may seem strange and even flat that Faust should find his highest happiness in a more or less prosaic engineering project."[2] It is true that today we are not as enthused about reclaiming land from swamps as was Goethe in the early nineteenth century. Yet Faust's soul was saved, not because he reclaimed land, but because, in Goethe's words, "whoever aspiring, struggles on, for him there is salvation." In this sense—in the knowledge that we are engaged in the struggle to improve the lot of Everyman—we can still share Goethe's enthusiasm, and a taste of Faust's salvation.

Our poets and novelists long ago stopped eulogizing engineering as a source of salvation, or even happiness, for mankind. But in the so-called underdeveloped lands, the gratification of helping people through technology is still expressed in art with unabashed enthusiasm. Take, for example, these verses from a contemporary Chinese poem:

The stars in the sky are crowded as sesame seeds,
The lights in the commune shine on our family.
Mama carries little brother in her lap,
Under the light she learns reading and writing.

Stars face stars in the sky,
Under the beam of every house hangs an electric lamp.
The stars are dim against the lamplight,
The electric lights shine in everybody's heart.[3]

How naïve such verse seems to us in our overlit, over-air-conditioned, overpowered homes. Yet it is naïve like the primitive painting of a Grandma Moses, which means that it is in some subtle way extremely refreshing. It mingles with the old Whitmanesque hymns of optimism that still reverberate deep within the engineer. As he busies himself with projects in environmental, agricultural, biomedical engineering, and the many other disciplines on which the welfare of mankind seem to depend, the ancient joy of helping the tribe to survive is constantly rekindled in the engineer's heart. Since the needs will always exist, the pleasures inherent in striving to meet these needs will persist as well.

In *The Bridge on the Drina*, Nobel prize-winning Yugoslavian novelist, Ivo Andric, retells an ancient myth in which the devil, envious of God's gift of the earth to man, scratched the earth's surface with his nails, seeking to prevent men from getting to and from their fields, from gathering together, in effect from surviving:

> When the angels saw how unfortunate men could not pass those abysses and ravines to finish the work they had to do, but tormented themselves and looked in vain and shouted from one side to the other, they spread their wings above those places and men were able to cross. So men learned from the angels of God how to build bridges, and therefore, after fountains, the greatest blessing is to build a bridge. . . .

The devil is still busy devising threats to our survival. (Often, as we know so well, the devil is within us.) So the gratification of learning from the angels how to ease man's lot is inherent in the human condition. To do such work continues to be, in Andric's words, "a blessing," and it is felt as such by the engineer.

If by any chance the day should arrive when the basic bodily needs of all men are fulfilled—and how unlikely that prospect appears at this moment!—the engineer's opportunity to contribute to the welfare of mankind will still be limitless. There will always be a need to plan for future development and maintenance of resources, to safeguard against the creation of new weapons, and to maximize the opportunities for people, not only to survive, but to pursue their interests in the fields of education, art, and recreation. Should the time ever arrive for the many, as it has for the few, when the greatest problem will become not poverty, but ennui, then the engineer would be called upon to assist in keeping alive the precious flame of enthusiasm. Such undertakings as the exploration of space, of the interior of the earth, and of the depths of the oceans would then be seen, more clearly than they are now, as fulfilling mankind's need for adventure and the lust for more knowledge. In such a society, engineers would also have more time and funds to aid archeologists, anthropologists, and art historians in their endlessly exciting quests for the treasures of man's past.

Even in an age of global affluence, the main existential pleasure of the engineer will always be to contribute to the well-being of his fellow man. It may occur to him that the satisfaction he finds in his work places him in the role of worker bee, and that perhaps the drones are having more "fun" than he. But the answer to these doubts is to be found in the concept of *maturity*. In all societies—one might almost say in all species—the child is free to play, and no one begrudges him his freedom. Indeed, he is often lovingly indulged. But, although there is a little bit of Peter Pan in each of us, maturity brings with it the desire to contribute to the communal welfare. The fulfillment of this yearning, I repeat, provides the engineer with his primary existential pleasure.

In making the point that existential pleasure is to be found in engineering, and in quoting the poetry and prose that I have selected, I do not want to imply that the practice of engineering keeps one in a state of perpetual ecstasy. Nothing could be farther from the truth. There are moments of great elation which envelop an engineer. But such moments are rare. Indeed if they were not rare they would cease to be precious. The experience of the engineer demonstrates that the good life—what John Dewey called the "satisfactory" as opposed to the merely "satisfying"—is achieved by immersion in the material world, by engaging in activity which is often mundane. The search for perpetual ecstasy—as generations of experience should have proved by now—can lead only to frustration.

I suggest that our present state of anxiety and alienation comes not, as so many savants tell us, from our materialistic concerns, but largely from our attempt to escape the natural confines of human life. We suffer from seeking ineffable fulfillment in mystical realms to which we have no access—except *through* the material life our philosophers scorn.

Franz Kafka makes the bewildered protagonist of his allegorical novel, *The Castle*, search desperately for salvation. His every effort to obtain recognition from the castle "authorities" is thwarted. He is lost in a nightmare world. All he knows is that in order to be saved he must be confirmed in his appointment to the job of *land surveyor*. In the midst of cosmic nothingness, Kafka saw that salvation might be found by turning away from abstract religion and philosophy and returning to a less intellectualized brick-and-mortar existence. Thomas Mann commented that Kafka meant to show how we "arrive closer to God through leading a normal life," enjoying "the blessings of bourgeois society."[4]

The engineer does not find existential pleasure by seeking it frontally. It comes to him gratuitiously, seeping into him unawares. He does not arise in the morning and say, "Today I shall find happiness." Quite the contrary. He arises and says, "Today I will do the work that needs to be done, the work for which I have been trained, the work which I want to do because in doing it I feel challenged and alive." Then happiness arrives mysteriously as a byproduct of his effort.

It is a paradoxical truth that when happiness is "pursued" it eludes us. In *Psychotherapy and Existentialism*, Viktor E. Frankl has observed that "pleasure is primarily and normally not an aim but an effect, let us say a side effect, of the achievement of a task." André Gide has expressed the same thought in slightly different terms: "Man's happiness lies not in freedom but in his acceptance of a duty." Or, in the words of the eccentric but acutely perceptive existential protagonist of Dostoevsky's *Notes from Underground*, "perhaps the only good on earth to which mankind is striving lies in this incessant process of attaining, in other words, in life itself. . . ."

The thrust of my argument has been to show that engineering is an occupation that responds to our deepest impulses, and is rich in spiritual and sensual rewards. Of necessity I have been speaking of engineering in the abstract—an ideal, if you will. The objection can be raised that this idealized definition of engineering cannot be applied in any meaningful way to the great diversity of individuals who are called engineers. We have already noted that the approximately one million American engineers exhibit an enormous variety of professional specialties ranging from designing electronic circuits to building dams, from devising theoretical models for systems analysis to testing new plastics, from conceiving new means of utilizing solar energy to selling machine parts. There are solitary geniuses working on discoveries that will amaze the world; and there are thousands of quasi-designers seated like galley slaves in huge drafting rooms. There are teachers and deans, brilliant teams in "think tanks," advisers to presidents, titans of industry, rugged individualists heading their own consulting firms; and there are thousands of frustrated inspectors for government agencies, and checkers of quality control in factories. Yet I visualize this vast, motley group as being part of one great profession, and I see each and every engineer as having access to the profound experiences I have discussed.

Metaphorically, I think of an engineering project as the staging of a production at one of the great opera houses of the world. The opera itself has been created by a composer and a librettist. A conductor interprets the music, and a director conceives and supervises the performance. There are star singers, singers in the lesser roles, and there is a chorus. There is an orchestra, a prompter, a concertmaster. There is a designer of sets, a designer of costumes, a designer of lighting, technicians who execute these designs, and stagehands. There are members of the board of directors, administrators, both artistic and technical, managers, comptrollers, treasurers, and assistants and secretaries.

Some of these people are more "professional" than others. Some are more "creative." One or two may be geniuses. But at the magic moment when the curtain rises, a performance takes place which in a real sense is a creation of the many people who have been working in their varying capacities. Not only do they each deserve credit for the finished product, but they each *experience* the satisfaction of having participated in a great undertaking. Some of those in subordinate roles hope to play a more important part in some future production, but whether or not this pertains, their share in the exhilaration of the performance is valid and earned. It may even happen that some of the lesser figures in the company will be more thrilled than those in the more important positions. Sensitivity and awareness are not necessarily proportionate to status. (By this I mean to say that it is entirely possible for young engineers to be more attuned than their more prestigious elders to the glories of engineering.) And let us not forget the audience —the public—for whose pleasure, and at whose expense, the performance has been produced. I do not want to overwork the metaphor. I do think, however, that it helps to demonstrate the way in which a large, diverse group can share in a common objective, and in a common gratification.

The mention of opera brings to mind the image of the *prima donna.* In his euphoric mood, the existential engineer must be careful not to overstate his case. There always remains the risk of being carried away with self-satisfaction, and lapsing back into the fatuous overoptimism of engineering's early years. We engineers also run the risk of becoming so contented with our data and our machines that we allow our senses to become deadened instead of energized. I have tried to show that engineering offers the opportunity of existential fulfillment. But the engineer who will not open himself up to these opportunities, who will not *feel* them, may very well end up as the inauthentic, smoothed-down man that the antitechnologists accuse him of being.

Probably the most famous rock 'n' roll song of our time is "A Day in the Life," recorded by the Beatles in 1967. The final line of this song—a line that has provoked much comment and controversy—says, "I'd love to turn you on." Some people have felt that this refers to the use of drugs. But Paul McCartney of the Beatles has been quoted as saying it means "turning people on to the truth about themselves." It is in this sense that I would like to see all engineers turned on to the myriad wonders, the existential pleasures, of their own profession.

Of course, man cannot live by engineering alone. He is a loving, playful, wisdom-seeking, beauty-adoring creature, and, deprived of the animating spirit of art and philosophy, his life is barren and his greatest works are as naught. But to seek love, pleasure, wisdom, and beauty without having the solid roots in life which one achieves only by constructive *activity,* is to cast oneself adrift in the empty space of aimlessness. As Arnold Jacob Wolf wrote of Reich's *Greening of America,* commenting on the book's "failure" five years after its publication:

> It is not possible to be without doing. The Sabbath is not the week. . . . Aimlessness does not fit the world which God made. . . . I do not believe that the task of life is to avoid the week or turn it into an endless hootenanny or folksing. Our task is to find a way to work that is really working and not also

> self-destructive—really doing the hard things that the world needs and the self needs.[5]

There is a scene near the end of Nikos Kazantzakis's novel *Zorba the Greek* in which a cableway, which has been arduously constructed to transport timber from a mountaintop for use as shoring in a mine, is tested for the first time and collapses in a heap. The scene is a comic wonder, with splinters and sparks flying, and monks and peasants scurrying away in fear and confusion. Zorba's boss, who has financed the project, sees his dreams in ruins. But, to the irrepressible Zorba, the matter of most concern is that the sheep, which has been roasting in preparation for a meal of celebration, might become burned. The meat is found to be cooked to a turn, a cask of wine is opened, and the two men begin to eat and drink. Soon they are dancing, arms about each other's shoulders, in the Greek style, Zorba teaching his boss the steps:

> "Bravo! You're a wonder!" cried Zorba, clapping his hands to mark the beat. "Bravo, youngster! To hell with paper and ink! To hell with goods and profits! To hell with mines and workmen and monasteries! And now that you, my boy, can dance as well and have learnt my language, what shan't we be able to tell each other!"

A joyous scene, a wonderful moment in literature. But soon Zorba and his young boss must part. The cableway lies in ruins, and the mine will never be developed. The workmen, who really needed their jobs, are now unemployed, but will somehow survive. Life goes on. Wine and dancing and an unquenchable spirit are what count in the end. "LIVE LIFE AND ENJOY IT!" is the message printed on the cover of the paperback edition of the book, as if only in pagan celebration is enjoyment to be found.

I sometimes fancy that the story has an epilogue in which the cableway and mine are successfully developed by means of a creative engineering effort. Then, at the celebration dinner, with the wine flowing and the people dancing, a special exultation is felt by those engineers who did the job that needed doing. I fancy them "carried beyond pleasure," like the C. P. Snow character whose laboratory experiment succeeds, or feeling as if their "stomach were being tickled by a paper balloon filled with some special light gas," like the young man in *The Woman in the Dunes* when he obtains fresh water from the sand.

I would have the engineers join in the drinking and the dancing. It is very important that they do not avoid the party and rush off to resume their work. But I would have them recognize that they are twice blessed—that if wine and dance are manifestations of mankind's primordial, existential spirit, no less so is the professional practice of engineering.

ACKNOWLEDGMENTS

Portions of this book appeared originally in "The Existential Pleasures of Engineering", a paper which I presented at a meeting of the Division of Engineering of the New York Academy of Sciences December 11, 1968, and which subsequently appeared in the *Transactions* of the Academy in May 1969. The Academy has kindly consented to the use of this material in the present volume. Some of the ideas set forth here have previously been expressed in articles which I have written for *American Engineer, Civil Engineering, Consulting Engineer, Engineer, Engineering Education, Engineering Issues, Technology and Culture,* and in the volume *Civil Engineers in the World Around Us* published in 1974 by the American Society of Civil Engineers. I appreciate the encouragement received through the years from the editors of these journals.

I am thankful to my editor at St. Martin's Press, Thomas Dunne, for his genial confidence in the undertaking, and for his challenging and constructive criticism. The manuscript was also read by my wife and by Milton Rugoff, both of whom made good suggestions—not always accepted—for improving my prose and clarifying my ideas. Valuable research help came from Rebecca Bushnell and Henry Tylbor and also from John Soroka and his staff at the Engineering Societies Library in New York City.

For several years my sons, David and Jonathan, have never missed an opportunity to point out to me articles and books which they thought pertained to my particular field of interest, and which consistently turned out to be novel and interesting. This book has been dedicated to them with thanks as well as with love.

My wife, Judy, in addition to giving practical help, somehow got us through the countless evenings and weekends of my solitary writing with just the right combination of supportive silence and cheerful intrusion.

I doubt that my feelings about engineering would have developed along such positive lines were it not for twenty years of pleasant experiences with my colleagues at Kreisler Borg Florman Construction Company. I also owe a debt of gratitude to the Thayer School of Engineering at Dartmouth College, where I first learned that engineering is not intended to be isolated from the quest for the good life, and where, through the years, I have observed faculty and students working at the creative and humanistic frontiers of our profession.

NOTES

CHAPTER 1

Of the many books which deal with the history of the engineering profession, my favorites are: Edwin T. Layton, Jr., *The Revolt of the Engineers*, The Press of Case Western Reserve University, Cleveland, 1971; Raymond H. Merritt, *Engineering in American Society, 1850-1875*, The University Press of Kentucky, 1969; and Alan Trachtenberg, *Brooklyn Bridge, Fact and Symbol*, Oxford University Press, New York, 1965. I have consulted these books often in preparing this chapter. I have also drawn upon Daniel Bell's introduction to Thorstein Veblen's *The Engineers and the Price System*, Harcourt, Brace & World, Inc., New York, 1963.

1. Charles Hermany, "Address at the Thirty-Sixth Annual Convention," *Transactions*, American Society of Civil Engineers, December, 1904.
2. Colonel H. G. Prout, "Some Relations of the Engineer to Society," in *Addresses to Engineering Students*, Waddell & Harrington, Kansas City, Mo., 1912.
3. George S. Morison, "Address at the Annual Convention," *Transactions*, American Society of Civil Engineers, June, 1895.
4. James Kip Finch, *Engineering and Western Civilization*, McGraw-Hill Book Company, Inc., New York, 1951, p. 321.

CHAPTER 2

1. Cat Stevens, "Where d' th' ch'ldr'n play?" Copyright 1970 Freshwater Music Ltd. (England) Controlled in the Western Hemisphere by Irving Music Inc. (B.M.I.) All Rights Reserved.
2. Tom Lehrer (first verse quoted in *This Fabulous Century 1950-1960*, Time Life Books, New York, 1970, p. 31; second verse from record, *That Was the Year that Was*, Words and Music by Tom Lehrer, All Songs ASCAP.)

CHAPTER 3

1. George M. Newcombe to the Joint Meeting of the Engineering and Scientific Manpower Commissions, May 22, 1969, New York; *Proceedings*, "Are Engineering and Science Relevant to Moral Issues in a Technological Society?"

2. For statistical information I have consulted *A Profile of the Engineering Profession*, A Report from the 1969 National Engineers Register, produced by Engineers Joint Council under a contract with the National Science Foundation. A check with the Council in 1975 confirmed that in general outline the "profile" does not appear to be changing appreciably. Figures from the Federal Bureau of Labor Statistics are not significantly different, taking into account the fact that government agencies classify people by job title or self-description without regard to educational background. Of the approximately $1\frac{1}{4}$ million Americans who called themselves engineers in 1970 and the $1\frac{1}{2}$ million estimated for 1980, less than two-thirds are "real" engineers as I would define the term, that is, professionals with engineering college degrees. On the other hand, many scientists and other academically qualified graduates practice engineering; some of these call themselves engineers, and some do not. Also, there are quite a number of graduate engineers who go into other careers and do not classify themselves as engineers. Obviously, precision in these figures is hard to come by.

3. *The New York Times*, February 16, 1975.

4. *The New York Times*, December 15, 1974.

5. Theodore Levitt, *The Third Sector*, AMACOM, Division of American Management Associations, New York, 1973, p. 33.

6. Martin Lang, "Developing an Elite Corps of Young Public Works Engineers for Municipal Projects," *Professional Engineer*, October, 1974.

7. Text reproduced in *Professional Engineer*, November, 1974.

8. Ronald B. Veys, Hydrological Engineer, Nebraska Natural Resources Commission, *Engineering News-Record*, January 23, 1975.

9. Jacques Barzun, "The Misbehavioral Sciences" in Richard Thruelsen and John Kobler (eds.) *Adventures of the Mind*, Alfred A. Knopf, New York, 1959, p. 20.

CHAPTER 4

The five books considered in this chapter are listed below, along with the pages from which quotes are taken:

Jacques Ellul, *The Technological Society*, Alfred A. Knopf, New York, 1948: pp. 14, 135, 320, 399, 221, 220, 275, 321, 417, 380, 65-66, 192, 34, 37, xxx, 145, (also in Chapter 6: p. 328).

Lewis Mumford, *The Myth of the Machine: I. Technics and Human Development*, Harcourt, Brace & World, Inc., New York, 1967; *The Myth of the*

Machine: II. The Pentagon of Power, Harcourt Brace Jovanovich, Inc., New York, 1970; *Interpretations and Forecasts: 1922-1972*, Harcourt Brace Jovanovich, Inc., New York, 1973 (referred to below as III):
pp. II-i, II-283, II-352, II-322, II-324, II-166, II-268, II-270, II-383, I-37, I-76, II-334, III-290, I-160, I-241, I-273, II-3, I-284, III-290, II-333, III-346 (also in Chapter 5: p. I-257; and Chapter 6: pp. III-491, III-492, II-413, II-433).

René Dubos, *So Human an Animal*, Charles Scribner's Sons, New York, 1968: pp. 219, 229, 196, 231, 191, 10, 179, 170, 196, 16, 16, 160, 178, 48, 195, 181, 194, 213, 219, 214, (also in Chapter 6: pp. 179, 180, 196, 237)

Charles A. Reich, *The Greening of America*, Random House, Inc., New York, 1970, Bantam Edition, 1971:
pp. 177, 384, 328, 201, 3, 287, 329, 6, 166, 28, 41, 86, 189, 307, 328, 415, 161 (also in Chapter 5: p. 165; and Chapter 6: pp. 179, 85, 382, 376)

Theodore Roszak, *Where the Wasteland Ends*, Doubleday & Company, Inc., Garden City, New York, 1972:
pp. 32, 235, 45, 39, 257, 23, 96, 26, 7, 23, 118, 428, 414 (also in Chapter 5: p. 379; Chapter 6: pp. 467, 420, 25, 413, 426 and Chapter 9: p. 174)

CHAPTER 5

1. Daniel Callahan, *Proceedings of the Centennial Convocation of the Thayer School of Engineering at Dartmouth College*, held September 23-25, 1971, published by The Thayer School, Hanover, N.H., April 1972, p. 74.
2. Donald W. Shriver, Jr., *Technology and Culture*, Volume 13, No. 4, October 1972, p. 534.
3. Lewis Mumford, *Technics and Civilization*, Harcourt, Brace & World, Inc., 1934, Harbinger Books Edition, 1963, p. 6.
4. Harold L. Wilensky, "Work, Careers and Leisure Styles," *Harvard University Program on Technology and Society, 1964-1972, A Final Review*, p. 142.
5. Based on an analysis of Bureau of Labor Statistics for job categories likely to have "assemblyline features." Noted in *Work in America*, Report of a Special Task Force to the Secretary of H.E.W., The M.I.T. Press, Cambridge, Mass. p. 13.
6. *ibid* pp. 14-15.
7. Saul Bellow, "Machines and Storybooks," *Harpers Magazine*, August, 1974, p. 59 (a paraphrase by Bellow of a quote from Russian writer V.V. Rozanov.)
8. Irene Loviss, "A Survey of Popular Attitudes Toward Technology," *Harvard University Program on Technology and Society, 1964-1972, A Final Review*, p. 174.
9. "Doctors Study Treating of Ills Brought on by Stress," *The New York Times*, June 10, 1973.

10. J. Huizinga, *The Waning of the Middle Ages*, St. Martin's Press, Inc., New York, 1949, Anchor Books, 1954, p. 67.

Also see Notes under Chapter 4 for quotes by Mumford, Reich, and Roszak.

CHAPTER 6

See Notes under Chapter 4 for quotes by Ellul, Mumford, Dubos, Reich, and Roszak.

CHAPTER 7

1. Concering the personality characteristics of engineers I have relied most heavily on *A Profile of the Engineer: A Comprehensive Study of Research Relating to the Engineer*, prepared by the Research Department of Deutsch & Shea, Inc., issued October 1957 by Industrial Relations Newsletter, Inc. This work is the result of an analysis of more than 40 books, studies, and reports in the field. I have also utilized a 1962 report by New England Consultants, Inc. entitled *The Engineer Today* "based on a synthesis of 452 references from the literature;" a 1968 study by R. P. Loomba entitled *An Examination of the Engineering Profession*; and *Engineering Manpower Bulletin*, Number 24, published by the Engineers Joint Council in 1973.

2. *A Profile of the Engineer* (See Note 1), p. II-8.

3. *ibid* , p. I-3.

4. Raymond H. Merritt, *Engineering in American Society 1850-1875*, The University Press of Kentucky, 1969, p. 118.

5. R. P. Loomba, *An Examination of the Engineering Profession*, 1968, and other references in Note 1. I must confess that I found some dissenting information and a disturbingly dour view in *The Engineers and the Social System* edited by sociologists Robert Perrucci and Joel E. Gerstl, (John Wiley & Sons, Inc., New York, 1969). This volume sees the engineer's primary goal as "the security of a middle-class income and a respectable job." (p. 145).

6. *A Profile of the Engineer* (See Note 1) p. I-13.

7. Thomas Mann, "The Making of the Magic Mountain," *The Atlantic*, January, 1953.

8. The studies cited in Note 1 make it clear that most engineers get great personal satisfaction from their daily activities. Again, as in Note 5, there is a dissenting view to be found in *The Engineers and the Social System*. It is noted in that book that "the rewards practicing engineers derive from their profession appear to be extrinsic rather than intrinsic" (p. 145). Also, a study of some M.I.T. graduates reports 46% "not very satisfied" with their engineering career. However, this source admits that its findings might be attributable to "the higher expectations of M.I.T. graduates," and quotes other studies that find engineers 35% "very satisfied," 41% "satisfied," and 20% with "average satisfaction." (See *Future Directions for Engineering Education*, copyright 1975 by M.I.T., published by the American Society for Engineering Education, p. 20).

Part of the problem here seems to stem from a failure to differentiate between an engineer's feelings about his profession and about his particular job. Another part of the problem has been stated by the engineering educator who complained that some studies of engineers have been guilty of selective perception and "smack of a social scientist's bias." (William K. Le Bold, Assistant to the Dean of Engineering at Purdue University, paper presented at the annual meeting of the American Society for Engineering Education, June, 1967). My own review of the literature, while far from exhaustive, strongly confirms the views that I have expressed in the text.

9. Address of Hon. W. J. McAlpine on assuming the chair after his election as President of the Society, September 2, 1868, *Transactions* American Society of Civil Engineers, Volume 1, 1867-71.

CHAPTER 8

The translations of Homer used are: *The Iliad of Homer*, translated by Richmond Lattimore, The University of Chicago Press, Chicago, 1951, and *The Odyssey of Homer*, Translated by Richmond Lattimore, Harper & Row, New York, 1965. The Biblical passages are from the King James Version.

1. *Iliad,* Book 18, lines 474-477.
2. *Odyssey,* Book 5, lines 234-257.
3. *Odyssey,* Book 9, lines 125-130.
4. Robert A. Butler, "Curiosity in Monkeys," *Scientific American*, February, 1954.

CHAPTER 9

1. See Note No. 2, Chapter 3, for statistical references.

See Notes under Chapter 4 for quote by Roszak.

CHAPTER 10

1. See Note No.2, Chapter 3, for statistical references. If half of the engineers are involved with machines, and a quarter are in civil engineering or allied fields, the question naturally arises concerning the remaining 25 percent. They appear to be about equally divided between science-oriented engineering—chemical, physical and mathematical—and, at the other end of the professional spectrum, such activities as business administration and industrial management.
2. John Gould Fletcher, "The Windmills," in *The Criterion Book of Modern American Verse*, W. H. Auden, ed., Criterion Books, New York, 1956.
3. Harriet Monroe, "The Turbine," in Harriet Monroe and Alice Corbin Henderson, *The New Poetry*, The Macmillan Company, New York, 1934.

CHAPTER 11

1. See Note No.2, Chapter 3, for statistical references.
2. George Madison Priest, "Notes" to Goethe's *Faust*, Alfred A. Knopf, Inc., New York, 1941, p. 418.
3. "The Electric Lights Shine in Everybody's Heart," A Nursery Rhyme of Ts'ao Hsien, Anhwie; Hsu, Kai-uy, trans. and ed., *Twentieth Century Chinese Poetry*, Doubleday & Company, Inc., Garden City, N.Y. 1963, pp. 419-420.
4. Thomas Mann, "Homage," Introduction to Franz Kafka's *The Castle*, Alfred A. Knopf, Inc., New York, 1930, p. xii.
5. Arnold Jacob Wolf, Jewish Chaplain, Yale University, "Consciousness Four," *Yale Alumni Magazine*, November, 1974, p. 19.